THE ETV MODEL BOOK

ROBERT SCHLEICHER

THE ETV MODEL BOOK

CHILTON BOOK COMPANY RADNOR, PENNSYLVANIA

1 2 3 4 5 6 7 8 9 0   8 7 6 5 4 3 2 1 0 9

Contents

Chapter 1 **Science Fact & Fantasy** **1**
The History of ETV Model Building, 4
Flying Model Rockets, 7
Space Fantasy, 11
The Hobby Aspects of ETV Model Building, 12
Models on a Grand Scale, 15

Chapter 2 **Modeling Techniques** **19**
Assembling Plastic Kits, 21
Tools, 21
Glues & Cements, 23
Preparing Parts, 27
Fitting Parts, 31
Filling Seams, 33
Drilling Holes, 35

Chapter 3 **Painting Miniatures** **37**
Air Brushing, 37
Inexpensive Painting Tips, 39
Masking Techniques, 41
Details & Special Finishes, 44
Decal Application Techniques, 46
Weathering, 47
Safety, 50

Chapter 4 **Advanced Modeling Techniques** **52**
Advanced Modeling Materials, 52
Working with Wood & Card, 55
Working with Plastics, 57

Chapter 5 **Model Rocketry** **63**
Flying Model Rockets, 63
Rocket Engines, 66

Rocketry Safety Code, 67
Launching Systems, 69
Tracking Rocket Flights, 74

Chapter 6   Rocket Kits & Interspace Vehicles   **81**
Basic Kits, 81
Interspace Rockets, 86
Designing Your Own, 94
Historical Rockets, 95
Scale Models, 100
Flying Saucers, 104

Chapter 7   Intraspace Vehicles   **107**
Designing Custom Spaceships, 107
The Battle Corvette Nebula, 111
The Petal-Class Fighters, 117
The Battle Cruiser Passim and
    Battle Carrier Roo, 129
The Ziggurat, 129

Chapter 8   Ground Effect Vehicles   **139**
Hovercraft Models, 140
Galax IV's Force, 141

Chapter 9   Space Buggies   **146**
Radio Control Magic, 146
Wheeled Vehicles, 151
Tracked Vehicles, 152

Chapter 10   The Space Scene   **154**
The Creation of a Planet, 154
Black Lighting Effects, 160
Special Effects Illumination, 161

Chapter 11   Photography   **166**
Depth of Field, 166
Black Light Photography, 169
In-Flight Photographs, 171

Glossary   **177**
Sources of Supply   **180**

THE ETV MODEL BOOK

CHAPTER **1**

Science Fact & Fantasy

Two of man's most compelling fascinations are his speculations about what might yet be and his reflections on what has been. Fantasies about flying, particularly about flying into the unknown, are often part of such speculations and reflections. This book will help you change those two-dimensional images of past and future space vehicles you've seen in books, films, or in your own imagination into three-dimensional models. There are even ways to take those miniatures and inject "life" into them—to make them fly under real rocket or jet power or with ground-effect power that looks like true anti-gravity. You can learn how to control those vehicles with inexpensive but sophisticated radio control transmitters and receivers, too.

It's a bit difficult to categorize the vehicles that stand at and beyond the edge of today's technology, whether they're models or reality. Some of them certainly are rockets, but others utilize the common wheel or the ducted-fan ground-effect (hovercraft) principle. Most fantasy vehicles presume a more sophisticated form of propulsion. The only term that seems to encompass all of these vehicles is "extraterrestrial," that is, anything that moves on, off, or between the surfaces of any planets, moons, satellites, or asteroids.

The models described in these pages are almost as wide-ranging in their construction as in their imagined uses. They include everything from ready-made plastic toys and plastic, wood, and card kits to scratchbuilt models that include all three materials plus metal and special lighting effects. Many of the models are built just for display, but there are often kits for identical vehicles that will actu-

1

Fig. 1-1 A model kit, like this 1/48 scale masterpiece, can be made into something almost real with the kind of care Robert Angelo put into it. This Star Wars™ X-Wing Fighter™ is an MPC product of Fundimensions, CPG Products Corp.

Fig. 1-2 There are a great many model rocketry hobbyists who build and fly every imaginable kind of rocket model.

ally fly under rocket power or via ducted fans driven by model airplane engines. The space buggies include ready-built models with battery-powered electric motors and gas-powered two-stroke engines, all with some type of radio control operation.

Extraterrestrial vehicle models are, in fact, a major hobby category. This new category includes virtually all the basic aspects of the hobby model field. If you've been thinking about building a detailed plastic kit, but aircraft, tanks, or cars don't inspire you, then start out with one of the space vehicles. The photos on these pages show what can be done with the simple-to-assemble plastic kits and how to do it. If you'd like to work with wood or cardboard rather than "modern" plastics, then try one of the flying rocket kits from firms like Estes or Centuri. Most of the all-plastic kits are also available from one of these two. If you'd rather not build at all, but just paint and/or weather, then you can do that by starting with one of the rather sophisticated "toy" plastic models or one of the cast metal miniatures. You can decide whether to go for a working version of your miniature or just settle for a display model. Having all those choices is, by the way, one thing that helps make ETV model building a hobby rather than just a fad or passing fancy.

3

The History of ETV Model Building

It's interesting to note that space and fantasy vehicles have been around as long as automobiles, aircraft, or any other segment of the kit industry. ETVs go back to the days when firms like Strombecker made wood models of the B-17 Flying Fortress. In the mid-fifties that same firm had a whole series of plastic model kits for ETVs, including a 1/80-scale replica of the TWA Moonliner from Disneyland's Tomorrowland. There were balsa models of Buck Rogers rockets in the mid-thirties, about the same time the first HO-scale train layouts were being built in America and England. Also in the thirties and forties, die-cast toy cars like the Tootsietoy line included a number of rocket vehicles.

The development of plastic injection-molding machines during the fifties permitted the hobby industry to grow into the multi-million dollar business that it is today. Space and fantasy kits came

Fig. 1-3 The Midwest brand Axiflo engines use model airplane engines to produce jet-like power for scale models like this Heinkel He162 designed by Nick Ziroli for Midwest models. *Courtesy Midwest Models.*

Fig. 1-4 Mike Williams assembled and detailed this Heller brand kit for the World War II-era Messerschmitt Me163.

and went—much like '52 Mercury automobiles came and went—from the various hobby companies' kit lines. Strombecker had a complete series of Disney-designed kits in the middle fifties when talk of actual men on the moon was just beginning. The late fifties saw the introduction of a number of arms-race military rockets. Scale models of the Jupiter C, Thor, and Vanguard were available from firms like Revell, Adams (no longer in the plastic kit business), Hawk (now part of Testors), Monogram and Aurora (which sold its plastic kit business to Monogram).

The big boom in ETV kits came, of course, during the peak years of the man-in-space program. The best-selling hobby kits for many manufacturers during the 1968–1972 period were scale models of the Apollo, Saturn, Vostok, Vanguard, and Titan rockets and the various space and lunar vehicles. The fantastically successful "Star Trek" television series appeared about this time, along with AMT's kits for the Klingon Battle Cruiser and the U.S.S. Enterprise. This was also the era of "Voyage to the Bottom of the Sea," "Lost in

Space," and "Land of the Giants" TV shows and Aurora's plastic kits for the vehicles from these series. The Aurora kits for the space vehicles in the film *2001: A Space Odyssey* appeared too, but the sales of the kits didn't match the success of the motion picture. There had been a lull in space and fantasy kit introductions and sales in the early sixties; in the mid-seventies it happened again. But the models used in the motion picture *Star Wars* brought interest to its present level. Complex surface detailing, weathering, and other minutiae rekindled public interest in the model maker's art.

The moon buggy, various types of robot surface probes, and other similar vehicles have made extraterrestrial transportation of humans a reality of the seventies. The miniatures of actual lunar vehicles and fantasy robots for similar ground surface exploration constitute another segment of the plastic hobby kit industry. Automotive-oriented model builders can now replace their designs for futuristic cars with models of vehicles that might well be used for exploration on moons or planets with gravity and atmospheric conditions different from those on Earth. The NASA moon buggy may seem as outdated as those Buck Rogers-era rocket-shaped automobile designs of the thirties, but you can let your own imagination redesign it so it will look more like the Star Wars™ Land Speeder™ and R2-D2™.

The future has been with us for more than two decades in the form of the hovercraft or ground-effect vehicle. The hovercraft is essentially a helicopter with a shroud or duct surrounding the outer circular path of the blade tips. The duct directs the air downward in a more uniform manner, so a smaller set of blades (or a propeller) can be used to lift any given amount of weight. The hovercraft was designed as a form of ATV (all terrain vehicle) for military use on swamps and other ground that would be difficult to traverse with either wheeled, tracked, or floating vehicles. Hovercraft are used commercially as ferries in many parts of the world.

The Land Speeder in *Star Wars* is available from Kenner, and its design is very similar to the round nosed, bulbous sides of some early hovercraft proposals. The Land Speeder has some unspecified form of anti-gravity, with jetlike forward propulsion, but a larger scale replica would make a most interesting fantasy vehicle based on the hovercraft principle. Testors has a ready-to-fly (ready-to-hover?) model called the Galax IV that operates on the hovercraft principle for free-flight action. It would be relatively simple to adapt the Testors operating parts to a scratchbuilt replica of the Star Wars Land Speeder, using either the Kenner model or the plans in Ballantine Books' packet of *Star Wars Blueprints* and balsa wood.

Flying Model Rockets

The flying model rocket was created by Vern Estes, the founder of the company that bears his name, in the late fifties. Estes developed the current solid-fuel model rocket engine and, of more importance to the hobby's growth, he made machines to manufacture the engines quickly and safely. Estes can hardly be listed as the founder of model rocketry, however, because the very nature of the vehicles demands that designs for what are usually war machines be tested as models before full-scale production can begin.

The history of rockets is a long one. They predated guns as weapons, and the Chinese probably had rocket-propelled lances and fire bombs as early as 700 AD. Isaac Newton presented his theory of action and reaction, the basis for the physics behind rockets, in 1678. Every major European power had military rockets by around 1826.

The age of liquid-fueled rockets and manned space travel can trace its beginnings back to 1903 when the Russian scientist Konstantin Tsiolovsky proposed both concepts. Dr. Robert Goddard, an American, test flew the first liquid-fueled rocket in 1926. In 1928, Germany's Fritz Opel flew a rocket-powered glider to become the first man to fly a rocket. The German V-2 rocket, the granddaddy of most of today's designs, was flown in Germany under the direction of Dr. Werner Von Braun. In 1949 America used Von Braun's V-2 to boost a WAC Corporal to an altitude of 244 miles to become first in outer space. In 1957 the Russians were first with their orbiting

Fig. 1-5 The pioneer V-2 rocket is a favorite prototype for many model kits, including this Estes model that really does fly. *Courtesy Estes Industries.*

Fig. 1-6 Scale model rockets, both display and flying versions like this Estes brand Honest John, are just one phase of the hobby. *Courtesy Estes Industries.*

spacecraft Sputnik I. The series of "first-swappings" culminated with our putting a man on the moon in 1969.

The interest in model rockets has ebbed and flowed almost in complete accordance with the public's interest in space programs. Estes' business was booming by 1959 and Centuri entered the field as a competitor in 1962. As we've already mentioned, the hobby ceased to grow during the late sixties, but boomed once again in the early seventies. There were several model companies in the flying model rocket business by 1972, including such giants as Cox and MPC.

Interest in space vehicles ebbed once again, and today Centuri and Estes are the major producers of flying model rocket kits. Vashon offered rockets fueled by compressed Freon during the early seventies. Its line was purchased by Estes but, with the federal in-

Fig. 1-7 The Estes 1/10 scale flying model kit of the U.S. Army's Pershing is a wood and card kit. *Courtesy Estes Industries.*

junction against fluorocarbon aerosols, Estes withdrew the liquid-fueled rockets from the market. The Estes and Centuri rockets are designed to take the cartridge-style rocket engines in standard sizes. There are dozens of different versions of these sizes that offer varying rates of thrust and/or time of thrust. A third firm, Flight Systems, Inc., makes a line of slightly larger engines and matching large rockets, as well as the smaller types.

Space Fantasy

Rocket travel to unknown worlds has fascinated man since the days of the first Chinese fireworks in the tenth century. Leonardo da Vinci's rocket designs of the 1500s were more his serious predictions of future fact than fantasy, but his contemporaries thought otherwise. The first known work of science fiction was *The Man in the Moon*, written by Bishop Francis Godwin in 1638. Many of Edgar Allan Poe's books can be considered to be science fiction tales, but his 1835 work, *Hans Pfall: A Tale*, actually used a space vehicle—a balloon. Jules Verne is generally acknowledged to be the real pioneer in the subject of science fiction in general, and space travel in particular, with his *From The Earth To The Moon*, published in 1865.

Science fiction was just one area of interest covered by the various pulp magazines of the twenties. The first all-Sci-Fi pulp was probably Hugo Gemsback's *Amazing Stories* in 1926. Buck Rogers was the talk of the comics in the thirties. Science fiction came into its own after the Second World War with hundreds of books and dozens of comic books, magazines, and films. This type of technically possible fantasy established some of the finest fiction writers of the decade and made names like Bradbury, Asimov, and Clarke part of college literature classes.

The motion picture industry did its part to establish science fantasy as an acceptable subject. Unfortunately, most of the low-budget films are as bad as any low budget western. However, the 1929 German film *Woman in the Moon* was decades ahead of its time technically. And space travel was well portrayed in the 1936 British film *Things to Come. When Worlds Collide* (1951) and *Desti-*

Fig. 1-8 There are at least a dozen different kits for flying and display versions of the pioneer rocket plane: the North American X-15. *U.S. Air Force photo.*

nation Moon (1950), both produced by George Pal, pioneered the technical excellence and precise special effects that were to startle audiences once again twenty years later in *2001: A Space Odyssey* and *Star Wars*. The difference between then and now is that you can easily bring the three-dimensional vehicles of today's films into your home in the form of accurate scale model kits and toys.

The Hobby Aspects of ETV Model Building

Some of us fancy ourselves as great artists or playwrights, others feel the need to create miniature railroad empires or wrist-loads of gem-studded jewelry. The ETV modeler is relatively new to the hobbyist field, in spite of the existence of stationary and flying rockets and other fantasy vehicles for two full decades. For the popularity of this new hobby you can credit the special effects people who worked with John Dykstra's team of model builders on *Star*

Fig. 1-9 Centuri's line of flying model rocket kits includes these historical miniatures (l to r): Little Joe II, Saturn Ib, Saturn V Apollo, NASA Skylab, and Mercury Redstone. *Courtesy Centuri Engineering.*

Fig. 1-10 Monogram's 1/128 scale set of plastic U.S. Space Missiles included everything of interest up to the kit's introduction in the early sixties. It's now a collector's item even though most of the rockets are offered as flying kits by Estes and Centuri. *Courtesy Monogram Models.*

Wars. Before them, space was the realm of the pristine white NASA-emblazoned vehicles of fact and fantasy films and publications. Dykstra took the screenplay's concept of a collection of used armament to heart. All the good guys' space ships in the film really do look like they've been through the wars. Much of their innards are exposed because, so the story goes, decorative outer skins are too hard for the repair crews to work on. The attacking fleet of X-Wing Fighters™ has that used look that brings a surprising credibility to them. The plethora of fine surface details on the film's Death Star™ and Imperial Battlecruiser™ are equally believable substitutes for that antiseptic white look of vehicles in earlier films and actual NASA photos.

The weathering effects and surface details make every one of the *Star Wars* vehicles a fascinating prototype for a model. In fact, the filming made those vehicles look real enough to be considered prototypes from the real world of the future. Hobbyists didn't even consider making models of some of the vehicles which appeared in low budget films. *2001: A Space Odyssey* created enough realism for one to believe the vehicles were real, but they were just too complex and imposing to serve as prototypes for models.

The various vehicles from the television series "Star Trek" make fine models because they are truly unique designs and because the stories they took part in were believable. Conversely, the vehicles in "Space: 1999" almost made up for rather poor stories and mediocre special effects.

The cumulative assortment of war rockets from yesteryear, man-on-the-moon rockets and machines, Star Trek℠, Space: 1999, and Star Wars kits, and research data on vehicles never kitted, makes a fine basis for a hobby. The fact that you can make any of these vehicles fly or hover or roll or track, as appropriate, makes things even more interesting. You certainly don't need to be a master model maker or a machinist, but you can just make the models to enjoy weathering them, animating them, photographing them, or admiring their design in three dimensions.

Models on a Grand Scale

Part of the realism of a model stems from the fact that it is a precisely proportioned reduction of the real thing in every dimension. That's generally called "accuracy of scale." If a doorway in a spacecraft is 8 feet tall, for example, and you have a model with that same doorway measuring ¼ inch, then your model is 1/384 scale (¼ inch divided by 8 feet x 12 inches per foot = 1/384). Any scale-model people who would inhabit that space vehicle should also be 1/384 size, as should all the major dimensions of the vehicle.

Unfortunately, very few of the fantasy vehicle model makers specify what scale they might be, even though virtually all the kits (as opposed to ready-built toys) really are accurate scale models. Fortunately, plans exist for some of the vehicles and others have doors or cockpits scaled to a human figure. The Star Wars T.I.E.-Fighter™ and X-Wing Fighter™ kits offered by the MPC Division of Fundimensions, a CPG Corporation, are very close to 1/48 scale. The AMT brand kits for the individual Star Trek U.S.S. Enterprise and Klingon Battle Cruiser are very close to 1/635 scale, and the set of three small Star Trek ships is 1/2200 scale.

The scale of the model is an important consideration when you are adding details like decals, cockpit interiors, or weathering, because they must be appropriate for the humans or human-size aliens who use the vehicle. You'll also need to consider scale if you plan to build additional vehicles that might not be available as kits or if you want to build a small diorama for the vehicles to rest on. The 1/48 scale of the MPC-brand Star Wars vehicles is helpful because there are a number of plastic model aircraft and tank kits that include details and scale model people that can be used to superdetail the stock kits and/or to make dioramas the way Robert Angelo did for many of the photos in this book.

The triangular rulers used by architects are helpful in measuring such scale details. These rulers are marked with an inch to the foot scale so that the "¼ inch" markings really mean ¼ inch *times* 12 inches, or 1/48 scale in the terms used by most modelers. You'll still have to do the math for odd scales like our hypothetical 1/384 or that Star Trek kit scale of 1/635.

The research necessary to determine the scale of any kit or proposed scratchbuilt model you might be considering is usually just a simple matter of doing the math. If there are plans, like those published by Ballantine for the Star Trek and Star Wars vehicles, then the math is all you need. If only photos exist, then consider any "spacemen" to be 6 feet high and work out the scale and overall size to fit the vehicle's occupants.

If you really are making the model from raw materials, then you might want to fudge the scale a bit to fit the parts from some commercial kit. If, for example, you were to find a photo of some spe-

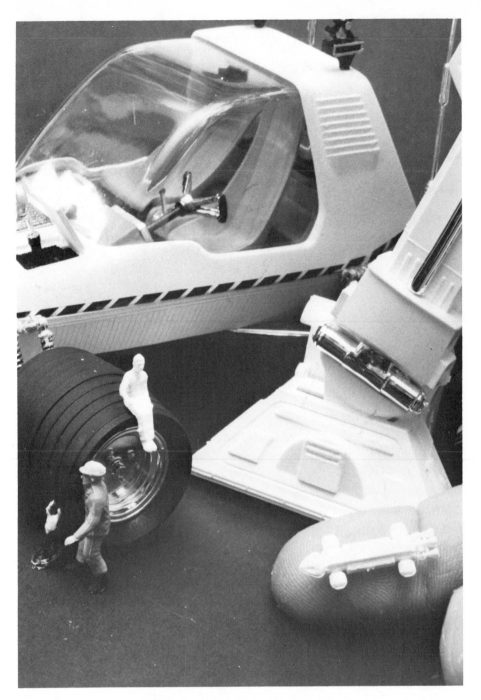

Fig. 1-11 ETV models range in scale from a fingertip size 1/800 scale Space:1999 Eagle to the 1/24 scale Space:1999 Alien moon buggie to the 1/6 scale MPC brand Star Wars™ R2-D2™ kit, of which one "foot" is visible. All models shown are MPC products of Fundimensions, CPG Products Corp. Both Space:1999 items are copyright © 1977 by ATV Licensing Limited.

Fig. 1-12 Two AIM-9L Sidewinder, two AIM-7F Sparrow missiles, and a collection of bombs and wing tanks suitable for kit conversion projects are included in the Minicraft/Hasegawa 1/32 scale F-16A kit. *Courtesy Minicraft Models.*

cial space vehicle from an old or new film that you determined to be 1/580 scale, then it would be wise to make a 1/635-scale scratchbuilt so you could use as many parts as possible from the AMT brand Star Trek series kits. Those MPC brand Star Wars models are probably closer to 1/44 scale than to 1/48, but the difference is not noticeable even in the hangar scene you'll see in Chapter 11.

There's a subliminal bonus to determining your ETV model's scale, too: once you decide on a scale, you have implied that the prototype vehicle really does exist! Your models, then, are not thought of as fantasy, but as replicas of a true full-size vehicle.

Science fiction and fantasy has been recognized as a true form of art and literature for the past two decades. The creation of what may come to be has its justifiable place of pride beside the creation of what is and what was. When you use a model kit or a scratchbuilt miniature to try to recreate what a vehicle looked like (or could look like) in three dimensions, you are truly practicing an art form. The model-building skills that created the 3-D vehicles that have been returned to two dimensions on these pages are the skills of artists in every sense of the word. The art of modeling extraterrestrial vehicles is the art of making fantasy seem reality.

CHAPTER **2**

Modeling Techniques

A well-finished model is as much a work of art as an oil painting, a ceramic piece, or a sculpture. The difference between a toy and a miniature work of art lies in the care with which the piece is finished. It's possible for an experienced modeler to turn a toy rocket into a work of art, but it does take some special skill because most toys do not have the correct proportions and most lack the details that are present in any hobby model kit.

Even a beginning modeler, however, can make a work of art out of any model kit. If you insist on thinking that a model kit is just a toy just because it's the same size as a toy, then *your* models will probably look like toys. The difference lies in the amount of care you are willing to put into the detailing of any particular model. You do, of course, need to know how to assemble, decal, paint, and weather—and that's just what I'll show you in this chapter and in the next.

I'd like to clear up a widespread misconception: the complexity of the kit has next to nothing to do with how "nice" or how realistic the model may be. Some of the simple plastic model cars that literally snap together with just a dozen pieces are far more realistic than some car models with 200 or more pieces, regardless of whether they're built by a beginner or by a professional model builder! Even some toys, when painted and finished with care, are far more true to life than many complex model kits. The important point is that the realism and artistic value of a completed model depends almost exclusively on how much care you put into it. The number of hours expended is totally irrelevant. There are models in

19

this book that were completed in 8 hours and are just as realistic as some that took 8 weeks!

You can decide just how much time you want to spend on any given model project. If you're a newcomer to the hobby, then start with as simple a kit as you can find and spend as much time as you have patience in assembling, painting, and decaling it properly. Then spend a few more hours applying a light touch of weathering or shading. I even recommend that you buy a plastic or metal toy miniature and use it to develop painting, decaling, and weathering techniques.

Avoid the complex *flying* model rocket kits until you have assembled at least one plastic kit and one simple flying rocket kit. Unfortunately, the most interesting flying model rocket kits are also the most complex and some are rather difficult to assemble. Spend those first few weeks or months learning the basic techniques of kit

Fig. 2-1 MPC brand 1/48 scale models built by Robert Angelo and photographed as described in Chapter 12. The Star Wars™ X-Wing Fighter™ (left) and T.I.E.-Fighter™ (right) are MPC products of Fundimensions, CPG Products Corp.

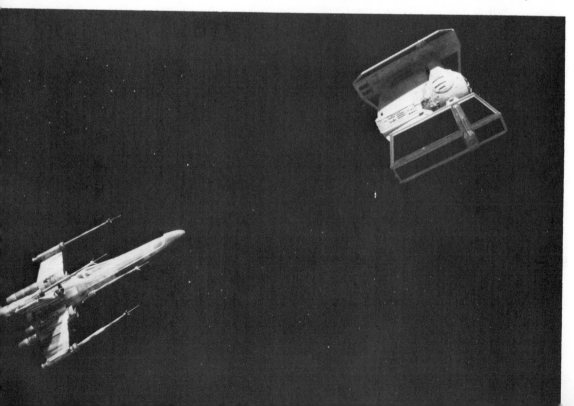

assembly, painting, decaling, and weathering. Then, if you find you enjoy the building part of the modeling hobby, go ahead and tackle one of those complex kits.

Assembling Plastic Kits

The actual assembly of the model is only about half of this hobby. If anything, the assembly is less important than the painting, decaling, and finishing phases of model building. I've seen many a sloppy glue joint, bits of mold flash, and even missing sections of the model disguised by a fine finish on the completed product. That's one reason I suggest that you consider using a toy or ready-built to develop painting, decaling, and weathering techniques.

An even better alternative is to consider that first kit as purely experimental or expendable. Try all the techniques you find in these pages and don't even think of keeping the model when you're done with it. That will take a lot of the pressure off of the learning process, and you'll be more likely to meet the challenge of applying paint and decals correctly. That practice model is the one to use as a learning tool for weathering, too, because you might not have the nerve to try it for the first time on a model you've finished to perfection.

Tools

There are only a few tools required to finish most plastic model kits, but the two most important items are the ones you'll most likely ignore: a sturdy, clean work area and plenty of light. A kitchen table or desk are the most common places to work, but both usually lack sufficient lighting. An old breadboard (or a new one—available for a few dollars at any lumberyard) makes a nice work area that will also help to prevent any damage to a table or desk.

Experienced plastic model builders often prefer to work on a sheet of $3/16$ or $1/4$-inch-thick clear Plexiglas-type plastic. Most model glues won't attack the plastic, so you can hold glue seams against it until the glue sets. The grain of the wooden breadboard can, in some rare cases, become permanently imprinted into the plastic model if liquid cement manages to flow between the model and the breadboard. So, if you do use a breadboard, tape a square foot or so of waxed paper to one corner and do all your gluing over the waxed paper.

Don't let the number of tools in the photo scare you. You can assemble just about any plastic kit with no more than an X-acto hobby knife, tweezers, and cement. The other tools will just make the job a bit easier and, perhaps, save some time. In some cases, there are several tools that do almost the same job, but one will work better for certain types of areas. I've divided the tools into categories, with the most important tool for each type of work at the top of the list and the others listed in approximate descending order of importance.

Cutting Tools
　　X-acto #1 knife with extra #11 blades
　　scissors
　　single-edge razor blades
　　X-acto #235 razor saw with #5 handle
　　diagonal wire cutters (preferably flush-cut type)
　　General Tool pin vise with #42, 51, 55, 70, and 80 bits
Holding Tools
　　pointed spring-open tweezers
　　masking tape
　　Scotch "Magic" tape
　　wood clothespins
　　pointed clamping tweezers
　　needlenose pliers
Filing Tools
　　#600 wet-or-dry sandpaper
　　round (needle) jewelers' file
　　flat (rectangular) jewelers' file
　　triangular jewelers' file
　　large medium-cut mill file

Fig. 2-2 The complete array of model building tools mentioned in the text. You can also get by with just tweezers and a hobby knife.

Measuring and Marking Tools
steel straightedge (General HO and 0-scale 12-inch ruler)
triangular architect's scale ruler
30–60 plastic drafting triangle

Some additional tools will be needed for painting, and you may want to purchase some of the tools shown in the next chapter for more advanced scratchbuilt projects. Any model in this book could, however, be assembled with only the tools listed above.

Glues and Cements

It takes more than just that typical tube of plastic model cement to build models as nice as those on these pages. It is almost impossible, in fact, to do a clean job of assembly with the tube-type cements. A bottle of liquid cement for plastics and some type of epoxy *must* be used to supplement the plastic cement. In fact, a bottle of liquid cement for plastic should be considered your standard glue for plastic models, with tube-type cements reserved only for a few special applications.

The tube-type cements for plastics include a certain amount of filler or dissolved plastic to make them thick. It is the solvent in the

23

cement, however, that does the actual gluing, and that solvent works the same whether it's pure solvent (in the liquid cements) or in thickened form (in the tube-type cements). The solvent dissolves the surface of the plastic at the joint and literally "welds" the surfaces together with dissolved plastic. Since the tube-type cements contain an amount of dissolved plastic, the actual glue joint is a mixture of the plastic from the two surfaces being joined and the plastic that has been dissolved in the glue itself.

The tube-type cements can take several years (yes, years!) to dry completely, particularly if an excessive amount of cement is used to fill in a badly fitted seam. Further, the tube-type cement often creates a joint that is soft and porous when compared to a similar joint made with liquid cements.

The one place where tube-type cement is essential is along those long seams like the edges of two fuselage halves on the AMT brand Star Trek models. The liquid cements will dry before they're

Fig. 2-3 An assortment of glues, including tube-type and liquid cement for plastics, cyanoacrylate cement, and five minute epoxy.

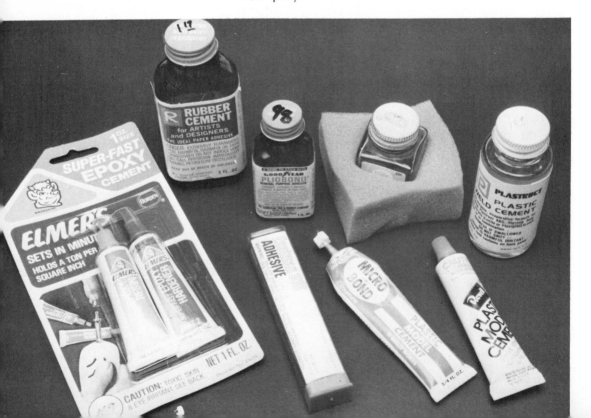

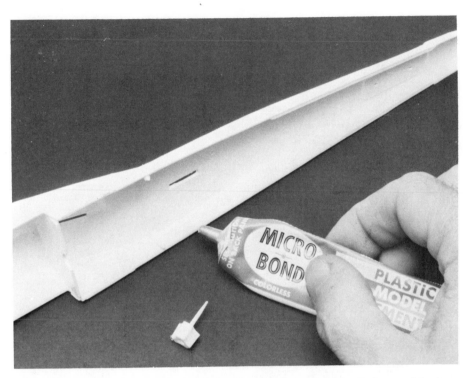

Fig. 2-4 The tube-type cements should only be used for long seams where the liquid would dry too quickly.

even spread along that length of seam. The trick to using the tube-type cements is to spread as thin a *continuous* bead as possible and to keep that bead toward the interior of the seam so glue won't ooze out when the two parts are pressed together.

Liquid cements for plastics are simply bottles of solvent that will dissolve the type of plastic used for most brands of hobby kits. The liquid cement should be applied with a brush. Some manufacturers furnish a brush right in the bottlecap. If yours did not, you may be able to find an empty bottle of fingernail polish remover or an empty bottle at a drug store with a brush in the cap to use for your liquid cement. Incidentally, do not try to substitute fingernail polish remover or any other type of solvent for genuine cement for plas-

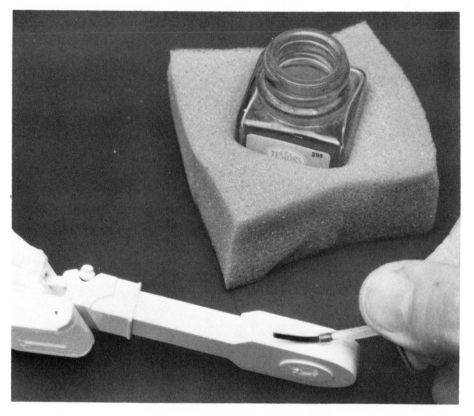

Fig. 2-5 Hold the parts together and apply liquid cement for plastics to the seam. A piece of foam rubber will help make a no-spill guard for the bottle. Cut a hole in the center of the foam with a knife.

tics. If you cannot find a bottle with a built-in brush in the lid, then just use a small paintbrush to apply the cement.

Brush a layer of the cement on the two surfaces to be joined and press them together immediately. You must work quickly with the liquid cement because it evaporates rapidly. If you have pressed the parts together and find that only a portion of the seam is actually cemented, you can apply additional cement to the joint and capillary action will draw it into the seam. Simply squeeze the parts together for about a minute and the joint should be self-holding.

One of the major advantages to the use of liquid cement is that you can work almost as quickly as you can apply the cement, with very little waiting for the cement to dry. The major disadvantage of using the liquid cement is that the parts must fit together snugly.

Since plastic cement only works by dissolving plastic, it's obvious that it will not work on chrome or painted surfaces. You must scrape the chrome away from the surfaces to be joined, or, if you have prepainted some of the small parts, scrape the paint away before gluing the parts to the model. You can use epoxy, rubber cement, Pliobond, or any of the various cyanoacrylate cements (Eastman 910, Zap, Hot Stuff, etc.,) to attach painted or chrome parts to the plastic, but the method of scraping, then using plastic cement, will make a stronger joint.

Do *not* use *any* type of plastic cement to attach clear plastic windows to a model. The plastic cements will craze and cloud the clear parts. Rubber cement or Pliobond will also attack most clear plastic surfaces.

The best glues for attaching clear plastic parts are the five-minute epoxies (I prefer Elmer's but there are dozens of alternatives) or the cyanoacrylates. The parts must fit together perfectly for the cyanoacrylates to work well because they require the absence of air to make their bond. The five-minute epoxies, white glues, and cyanoacrylates are also good to use when attaching metal, wood, paper, cloth, wire, or thread to a plastic surface. There is really no place in modeling for various paste glues or for all-purpose household glues; the adhesives I've discussed in this section will do the job so much more effectively.

Preparing Parts

One would think that the proper assortment of tools and the right choice of glues would be all that's needed to get to building. But there's more to it than that. Actually, the first step in assembling *any* kit is that seldom-practiced (and, therefore, often mysterious) art called "understanding the instructions." Reading the instructions is not enough; you must understand the way they want you to

Fig. 2-6 The flush-cut diagonal cutters can be used to cut the parts from the sprue as shown. Do not use conventional cutters here.

assemble the kit. (While you're building a basic understanding of how the kit is supposed to go together, decide what colors you'll want to paint the various pieces. You may want to paint some of the tiny parts while they are still attached to the molding sprues.) Only after you thoroughly understand the way the manufacturer made the kit to fit do you remove any of the major pieces from their molding sprues and test-fit them. Prior to assembly, each and every plastic part must receive some preparation and full inspection as well as a test fit with its adjoining parts. That's right; the parts in those plastic kits are *not* ready to assemble—not if you expect to have a model rather than a plastic toy, that is.

All of the small parts will still be attached to the molding sprues which are, in fact, the insides of the pipes or cavities that carried the molten plastic into the mold to make the actual pieces. The parts must be cut—never broken away—from those molding sprues. If you try to break the part off, there's at least a fifty-fifty chance you'll break off a small chunk of the part and leave it on the sprue. You can buy what are called flush-cut diagonal cutters (dikes) at many model railroad and electronics hobby stores. These dikes make a wedge-shaped cut, but the angle on one side on the wedge is 180 degrees. Once you learn to hold them with the flush-cut side to-

Fig. 2-7 The parts can also be cut from the sprue with a hobby knife. Scrape any plating from the joining surfaces of plated parts.

ward the part, you can snip the part from the sprues in a hurry and damage *only* the sprue in the process.

There are a few other things that you should look for while you're test-fitting the kit's pieces, namely mold marks, parting lines, and ejection pin marks. The first term includes the other two. The parting lines, sometimes called "flash," are the places where the molten plastic tried to squeeze between the two halves of the mold when the part was being formed. All traces of parting lines must be removed from the model with a hobby knife to assure proper fit of the parts and to eliminate unwanted seams and lines. It is sometimes difficult to tell a parting line from one of the molded-in detail lines, but you can do it if you realize that some trace of the parting line will travel completely around the part. There may be more than one parting line, too.

Fig. 2-8 The sharp edges on these parts are examples of flash that should be removed before assembling the parts.

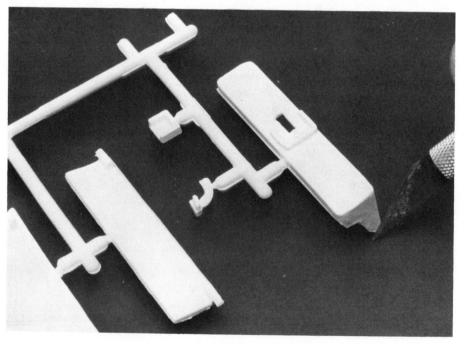

In some instances there may be a wisp of plastic blocking an assembly or alignment hole. The round jeweler's file is the tool to use to clean out those holes. The rectangular and triangular jeweler's files can make it much easier to remove that flash and those parting lines from inside windows and other close-tolerance areas.

The ejection pin marks are round or half-round marks between $1/16$ inch and $1/4$ inch in diameter that appear where steel pins in the molds were used to force the plastic parts out of the mold. The ejection pin marks will usually appear on the inside surfaces of the model where they don't cause any problems. Sometimes, though, the marks will interfere with the fit of some of the interior detail pieces and/or clear plastic window pieces. They may also appear along the joining seams of some parts and create an irregular seam once the model is assembled. You should look for any bothersome ejection pin marks while you're test fitting all the pieces. Remove them by trimming them flush with the model's surface with a hobby knife—just be sure to remove only the offending portion of the plastic and not part of the model.

Take a few more minutes, before you actually begin to apply cement to those seams, to decide on how you are going to paint the model. If you are going to paint it two colors with the color separation line along a seam between two parts, you may want to assemble the model in subassemblies and paint it before final assembly. Generaly, the parts don't fit together well enough to give you the luxury of pre-painting; joints that need pre-fitting and careful gluing would ruin any such efforts. You can do whatever seems easiest to you. I assemble everything but working parts and clear plastic parts (assuming the clear plastic *can* be added after the model is assembled), apply a coat of primer, a coat of the primary color, and then touch-up the detail colors with a paint brush.

Fitting Parts

Fitting parts properly is critical to the strength and appearance of the finished model. The most common mistake new modelers make is to use cement to try to fill in seams and joints that don't fit.

You can't do it with liquid cement, so these newcomers resort to great globs of tube-type cement. Those great globs of cement will continue to shrink *after* the model is painted—sinks or potholes will appear at random along the glue joint and continue to deepen for months as the remaining solvent evaporates.

It will probably be necessary to hold the parts together with masking tape or clothespins, even with a joint that fits well, because the plastic itself may be slightly warped. The taped or clamped joint is the better alternative to the use of too much glue. If you use masking tape, place a pair of toothpicks, one on either side of the seam and parallel to it, to hold the tape away from the seam itself. If

Fig. 2-9 Use masking tape and clothespins to hold parts together until the glue dries.

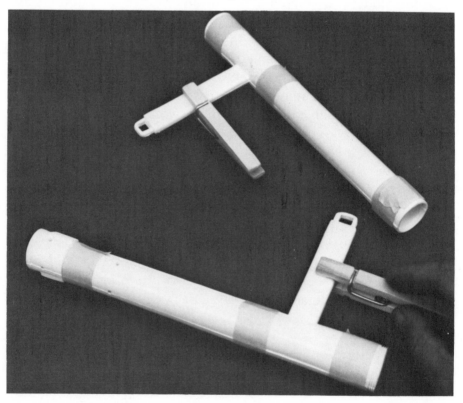

you allow the tape to touch the seam, the plastic cement will find its way out of the seam and beneath the tape. The same is true when using rubber bands instead of masking tape. A pair of toothpicks positioned at each side of the seam where the rubber band crosses it will keep the glue where it belongs.

Filling Seams

There are some models that will have glaring glue seams no matter how well you apply the cement or how tightly you clamp the joints together. One of the reasons I recommend using primer before the final coat of paint is so you can see those ugly seams before it's too late to do anything about it. If you know that there's a problem area before you assemble the kit, you can sometimes solve the problem by using a lot of pressure on the seam while the glue is still wet, along with a half dozen or more applications of liquid

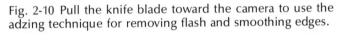

Fig. 2-10 Pull the knife blade toward the camera to use the adzing technique for removing flash and smoothing edges.

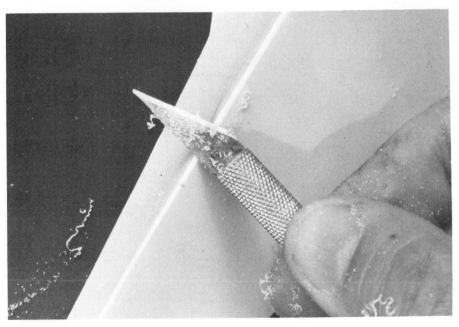

cement for plastics. The liquid cement and the pressure will cause a thin bead of the dissolved plastic to ooze out of the seam. If you can clamp the parts together without touching that bead of cement and plastic until it has dried for 48 hours or so, you may then be able to neatly trim the bead away with a hobby knife to leave an invisible seam that's ready to paint. Sometimes scraping the seams with the backwards stroke of a hobby knife (called "adzing" in the woodworking trade) will remove just enough plastic from the seam and push just enough of it into the seam to hide it completely once primer and paint are applied.

Some seams are just too wide and too deep to be hidden dur-

Fig. 2-11 Automobile body spot putty (for patches) works well in filling seams and other blemishes in plastic or wood models.

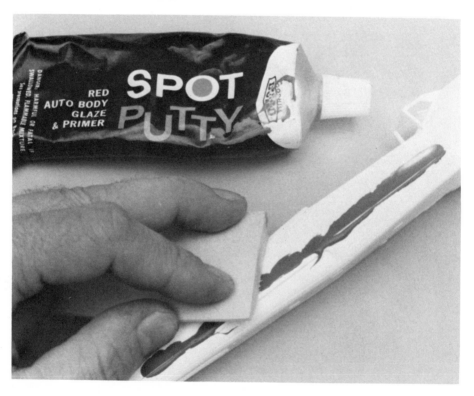

ing the assembly process. Those seams are best filled by applying some automobile body touch-up or spot putty. The thin automobile body putty that is intended for touch-up or spot repairs is the best—the thicker putties often leave air pockets and rougher surfaces. There's no need for the two part epoxy-like putties for simple seam filling jobs.

Spread the putty over the seam with the tip of a screwdriver and apply enough putty so it extends upward above the surface of the model a few thousandths of an inch. Be careful to spread the putty only near the seam so you don't have to sand away adjacent details to remove the putty. If there are panel lines that cross the seam, these can be scribed back into the surface of the plastic with the hobby knife.

Let the putty dry overnight and then use number 600 wet-or-dry sandpaper wrapped around a block of wood to sand the putty level with the surface of the plastic. Spray on a light coat of primer and check to see if the seam has vanished. Some models may require two or three applications of putty and primer to remove the seam lines.

Drilling Holes

There should be no need to drill any holes in most plastic hobby kits. You may find, though, that you have accidentally filled in a hole with body putty or that you need a hole to attach a wire antenna or some other detail. You will probably need to drill several holes if you want to scratchbuild your own models. Model railroad shops and some large hardware stores carry a small hand drill called a "pin vise." The one made by General Tool seems to be the easiest to use and it comes with four different collets to allow the use of almost any drill from a hair size number 80 to ⅛ inch.

The pin vise will also allow you to use model size screws like a 00–90, 0–80, and 2–56 to assemble some parts, for pivot points on landing gear and other moving parts, or just to provide hex head or round head surface details. The plastic used in model kits is soft enough so that these small screws will be self-tapping so all you

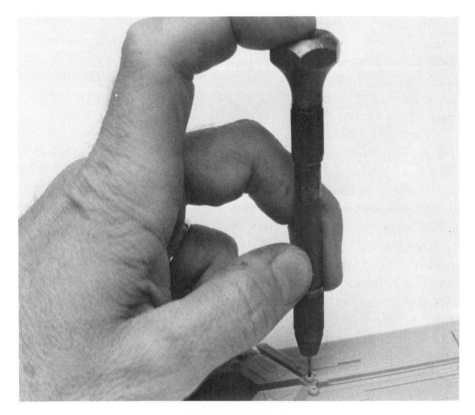

Fig. 2-12 The General Tool pin vise should be held as shown so the body can be rotated with your center finger and thumb while keeping downward pressure on the drill with your index finger.

need is a tap size hole (a #63 for a 00–90, #55 for a 0–80, or #50 for a 2–56 screw). The store that sold you the pin vise should also be able to supply the numbered drill bits and the small screws, hex-head bolts, and nuts.

CHAPTER **3**

Painting Miniatures

The painting procedures and materials are essentially the same for plastic model finishing as they are for finishing models made of wood, card, or metal. The critical coat of paint for plastics is the primer. The primer must completely cover the model with a very thin coat of paint *and* not attack or etch the plastic itself. Most primers sold in hardware stores produce a coat so thick it hides many details on the model. The Tempo brand primer and the Magic brand (sold by Standard Brands paint stores) are two that seem to work quite well. If you cannot find those, then you'll have to experiment on some leftover plastic sprues or scraps until you find a brand that covers with a thin coat and does not eat into the plastic.

The two primers I suggested are sold in aerosol cans to make their application easy. Don't even try brush on primers because you'll have to use too thick a coat to cover the model completely. If you have an air brush, you can use the canned or bottled primers, including the fine primers intended for use on automobiles. The kind that can be thinned with lacquer thinner are usually safe for use on plastic surfaces.

The finish or color coats of paint can attack some brands of primer. If you are using a lacquer or Floquil brand paints for those final coats of color, then you may have to use a primer from the same firm that makes the paint. Floquil makes a fluid called "Barrier" to allow the use of their paint and primer on plastic.

Air Brushing

An air brush is next in importance to cement, tweezers, and a hobby knife in what I would suggest as tools essential for creating

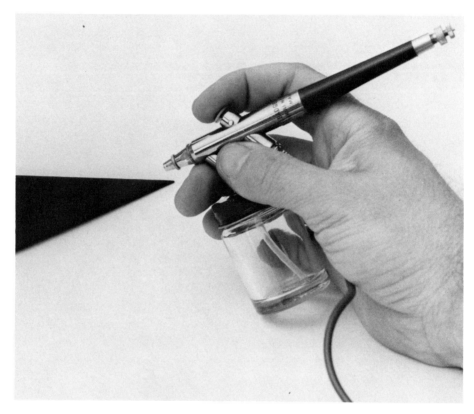

Fig. 3-1 The Badger model 200 double action air brush being used to paint one of the Petal-Class Fighters. The air compressor for the air brush is out of sight at the end of the supply hose.

realistic miniatures. Unfortunately, the air brush is one of the most expensive tools in the hobby because, to be effective, it must be used with an air compressor or air tank that costs $80 or more plus the $20 to $60 for the air brush itself. The air brush is just a miniature version of the spray guns that are used to paint full size automobiles. A single-action air brush is fine for model work, but it should have internal mixing to be worth the investment. The Badger model 350 and model 200 air brushes and the Thayer & Chandler model FP are popular with modelers. Air compressors are available at Sears and most hardware stores. Pick one that supplies a minimum of 25 pounds per square inch (psi) of air pressure and at least one cubic foot per minute of air flow. It's best to spend another $30 and get an air trap and air regulator with gauge like Badger's

That $130 to $170 investment in an air brush will enable you to spray most brands of model paint right on the surface of the model, to mix your own colors, to apply just enough paint to color the model without hiding the finest details, and to produce the most realistic types of weathering and shading effects. You can do *almost* as good a job with a paint brush and sponge. In this book, the various MPC brand models of Star Wars vehicles were weathered with an air brush, and the Battle Corvette Nebula shown in chapter 7 was weathered with a spray can and a sponge. You can compare them and decide if it's worth the investment to you to purchase an air brush. Don't bother to buy any of the less expensive air brushes because all they really can do is let you mix or thin your own paint. Their spray action is virtually the same as that of most aerosol cans of spray paint. The better air brushes will allow you to adjust the spray pattern so fine you can literally spray a dot as small as the period at the end of this sentence.

The air brush will have instructions for its operation and cleaning. Cleaning is particularly important—the air brush must be cleaned thoroughly *every* time you use it.

For use with the air brush, I recommend Scalecoat model railroad paint, if you can find it, or Floquil's paints. Thin the paint with about an equal amount of thinner and use about 25 psi of pressure with the air brush held 6 to 12 inches from the model. Adjust the paint flow at the air brush to give the paint spray pattern you want. For weathering and detail work, slightly less air pressure and a finer spray pattern should be used.

Inexpensive Painting Tips

Frankly, an air brush is a luxury that few modelers can afford and it really is a luxury no matter how enjoyable. You can achieve virtually the same results by using aerosol cans for the principal color coats, a brush for fine details, and a sponge for weathering.

I cannot recommend the use of most of the spray paints sold by hardware and paint stores because most, even those that are flat or non-gloss colors, are too thick for use on models. Floquil and Pactra, however, make a number of flat finish colors that are perfect for space style vehicles. (Although decals will adhere better to a glossy (shiny) finish, there are ways around that problem.) The flat finish paints seem to cover with a thinner coat than gloss paints. Even with an extra coat of gloss prior to decal application, the flat or non-gloss paints seem to work best for most models. Floquil, Pactra, Scalecoat, Polly S, most brands of acrylic paint, or Testors bottled flat finish paints can be applied with a brush, but that's often difficult thanks to the complex surfaces and shapes of ETV models.

The Floquil paints must be applied over a primer coat of Floquil's Barrier (also available in aerosol cans), or else the Floquil

Fig. 3-2 The best paints to use are those designed expressly for models. Use brands like Pactra, Scalecoat, Testors, Floquil, or Polly S, and a suitable primer paint like Magic brand.

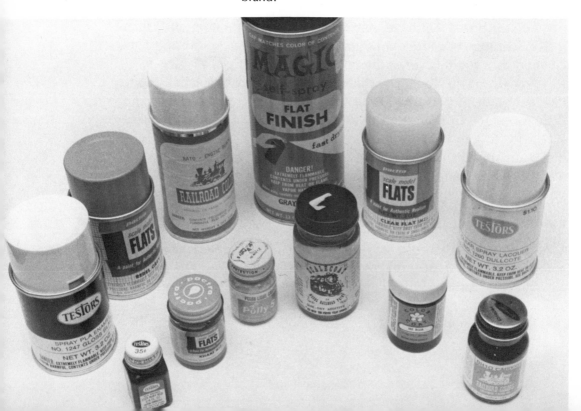

may crack like the surface of a dry lake. Pactra's aerosol cans of flat finish paint can be used without a primer coat (assuming you have no cracks to be filled with the body putty and primer), so there's less chance of burying fine surface details under paint. Testors makes equally suitable aerosol cans of flat finish paints, but only in red, black, and white.

The painting and finishing procedure I recommend whether using a spray can or an air brush is as follows. Use a wire hook, bent from a coat hanger, to hold the model during all of the painting steps.

1. Wash the model in detergent and rinse in warm water, and then allow it to air dry (or use a hair dryer) so there's no lint accumulation.

2. Mask any clear windows or canopies with masking fluid or Scotch Magic tape.

3. Spray on a coat of Pactra or Testors flat-finish color (if it's a two color finish, spray on the lighter color first and then mask it and apply the darker color).

4. Spray on as light a coat as possible of Testors Glosscote or Pactra clear gloss.

5. Apply decals.

6. Spray on a coat of Testors Dullcote for a really flat finish or Pactra Clear Flat for a semi-gloss finish.

The weathering coats, if desired, can be applied before or after the final coat of clear depending on the effect you want to achieve.

Masking Techniques

You will need to use paint masking techniques to protect clear windows or canopies and to cover sections of the painted surface when adding a second color on a two-color model. Scotch Magic tape works much better than conventional masking tape because it is thinner (so the paint doesn't build up along its edge) and because you can actually see, by the disappearance of that cloudy look, when the tape is stuck firmly to the model. If you are masking a curved line or one with zig zags or steps, just apply too much tape

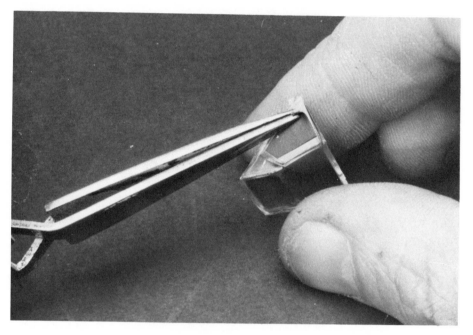

Fig. 3-3 This canopy was spray painted after being masked with Scotch Magic tape. Here the tape is being removed from the clear portion of the canopy.

and use a hobby knife to trim the edge of the color separation line you want. Spray on the paint, applying the thinnest possible coat of paint, especially along the tape's edge, and allow it to dry. If you can reach the edge of the tape, it's best to cut through the paint along the edge of the tape with a new hobby knife blade so there's no chance for the paint to stick to the tape. Now pull the tape back over itself. The paint should tear off evenly with the tape. If you find that the tape is pulling the first color with it, then you probably didn't let the first coat dry or you forgot to wash the model before painting it and a fingerprint or other grease smudge beneath the paint caused it not to adhere. Any ragged portions of the edge can be touched up with a dab of the darker color applied with a 00 paint brush.

There are some trick techniques for masking the clear portions

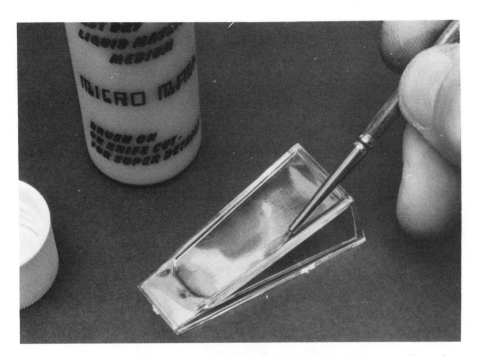

Fig. 3-4 Masking fluids are applied like paint, allowed to dry, then peeled off after the part has been painted.

of windows and canopies. The easiest of them is to cover the entire canopy or window with Scotch Magic tape. Use a hobby knife to lightly cut along all the edges of each clear window panel. Next, peel away the tape from the frame portion of the window. The window can then be spray-painted at the same time as the rest of the model. When the paint has dried, pick one edge of the tape with the hobby knife blade and peel the tape away to reveal the clear window areas.

The alternate technique utilizes one of the masking fluids sold by hobby and model railroad shops. Cary brand Magic Masker and MicroScale's Micro-Mask are two examples. This fluid brushes on like thick paint and dries to a rubbery consistency that is strong enough to be peeled from the surface of the model. The masking fluid is painted onto the clear portions of the windows or canopies and, when the fluid is dry, the windows or canopies are spray-painted. Next, trace the edges with a new hobby knife blade to slice through the paint and masking fluid. Finally, peel the masking fluid away from the clear areas.

Details & Special Finishes

Most of the detail work on an ETV model should be done with flat black, flat red, or flat blue paints. The blue and red can be used for accents and trim to contrast with the flat white or light grey that is the usual color choice for such vehicles. Flat black can be used to "open" gun barrels and viewing ports and to accent panel lines. Use one of the water base artists' acrylic paints and thin it with

Fig. 3-5 To accent panel seams for greater realism, slice along them with a hobby knife and then rub water-thinned acrylic paint over the cuts so the paint adheres only to the cuts.

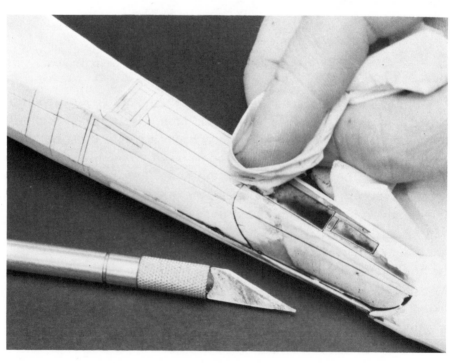

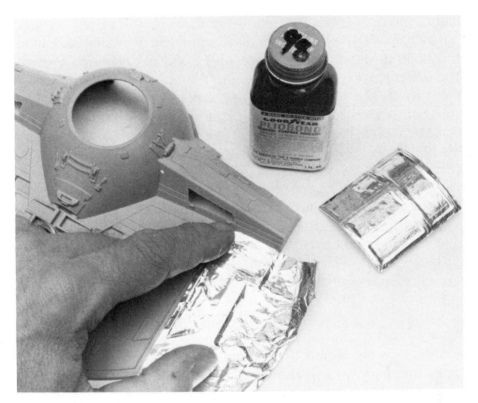

Fig. 3-6 Plain aluminum foil can be used for simulated-aluminum ETV models. Cut the foil into panels no larger than 1 x 2 inches and glue it to the model with Goodyear Pliobond cement.

about 99 parts water. Wipe this wash over the model so it sticks into the crevices and corners to give a highlighting effect to all the unblackened surfaces.

For extra detail, slice over all the panel lines with a hobby knife *before* rubbing the black wash on with a rag or sponge. The black wash will flow into those cut lines to make them more visible and realistic. Just be sure to wipe the black wash away before it has time to dry. It's best to work on just a few square inches of the model's surface at a time when accenting panel seams and highlights.

Aluminum surfaces look especially realistic on a space vehicle, particularly as a contrast to the white or grey that is normally used on such models. You can simulate aluminum by using aluminum foil glued to the model's surface with Goodyear Pliobond cement. The secret to using foil is to apply just a couple of square inches at a time. Pieces the size of one or two panels on the model's surface are about right. Apply an oversize piece and, when it is firmly in place, cut along the edges of the panel lines just inside the edges of the foil. Repeat the applications with additional pieces of foil. A final coat of Testors Dullcote or Pactra Clear Flat will maintain the foil's shine.

The best method for simulating an aluminum surface requires the use of an air brush. Liqu-A-Plate brand metal finish paints are sold only in bottles and they should be applied as a spray from an air brush. Liqu-A-Plate is available in several shades ranging from a silver-grey shade of aluminum to an almost black gunmetal shade. Individual panels or complete areas of the model can be painted in almost matching shades of Liqu-A-Plate for a most unusual effect. The best part of Liqu-A-Plate paint is that it can actually be polished by simply rubbing the dry painted surface with a dry rag. The patchwork effect of a modern airliner can be duplicated by spraying the entire model with aluminum Liqu-A-Plate and then polishing the surface. Mask a random selection of individual panels with Scotch Magic tape, then respray and repolish. The second coat and polish will produce a slightly darker surface that is incredibly realistic.

Decal Application Techniques

There is really only one right way to apply decals to any model. If you skip any one of the steps, you'll pass up the chance to have those decal markings look exactly like they were painted on. The first step in decal application is to spray the model with a thin coat of clear gloss and to let it dry overnight. Cut each decal as close to the printed portion as possible (with the Micro-Scale decals, cut the paper *near* the clear area) to remove as much of the clear area as possible. Dip each decal in water and set it aside on a blotter while the decal's glue is dissolving. From then on, handle the decals *only* with tweezers or the tip of a knife blade—never with your fingers.

When each decal can be moved on its paper backing, it is ready to be positioned on the model. Place both the decal and its paper backing where you want the decal. Hold the decal itself with the tip of a hobby knife blade while you pull the paper backing from beneath it. Apply a drop or two of one of the decal softening fluids (like Walthers' Solvaset, Champ's Decal Set, or Micro-Scale's Micro Sol) to the decal and let the fluid evaporate. The decal softening fluid will make the decal soft enough to snuggle tightly around curves, ridges, or cracks for that painted on look. It may take three or four applications of the decal softening fluid to get some brands of decals to snuggle tightly against complex surfaces. If the decal covers one of the model's simulated seams, you may have to slice through the decal to get it to adhere to the edges of the seam without any air pockets. A light dab with a finger wrapped tightly in Kleenex can help to push the decal down. If there are any trapped air bubbles (visible as cloudy dots on the decal's surface), prick them with the tip of the hobby knife blade and apply one more coat of decal softening fluid. Let the fluid dry overnight and gently scrub the model's surface with a damp cotton swab to remove any traces of decal glue or decal softening solution. Let the model dry overnight again. A final coat of clear flat (Testors Dullcote or Pactra's Clear Flat) will hide the slightly raised edge of the decal and protect it from being accidentally soaked or scratched away.

Weathering

A light touch of weathering will make even a pristine clean rocket like a Saturn look more realistic, and weathering is virtually an essential to the completion of most of the fantasy vehicles like Star Wars X-Wing Fighters. "Weathering" is used to describe the visible effects of rain, dust, meteor storms, exhaust gases, explosion clouds, laser created debris, and the like on the surface of a vehicle. Weathering takes two basic forms: dusty looking puff-like patterns, and various degrees of streaks. In an extreme case, weathering can even be applied to "battle damage" that can tear away part of the

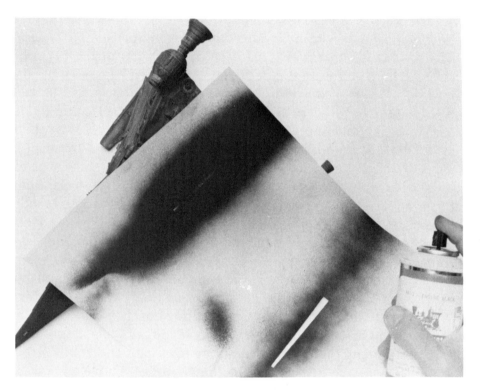

Fig. 3-7 Aerosol cans can do many of the jobs usually reserved for an air brush. Cut a paper mask to spray on exhaust streak stains.

vehicle's outer skin to reveal the framework or part of the engines.

Weathering is simulated on a model by applying a wash of light gray (to a dark surface), dark brown, or dark gray with a sponge or an air brush. The wash consists of about 95 to 99 parts thinner and the remaining parts paint. If you are using a sponge to apply weathering, I suggest using one of the water-base acrylic paints thinned with water. Regular Floquil or Scalecoat paints and thinner can be used with an air brush. You can also use a spray can with a paper mask to simulate the splatter effect of air brush applied weathering.

The basic weathering coat is almost a dip to deposit paint in the crevices and hollows of the model so the rest of it shows up more

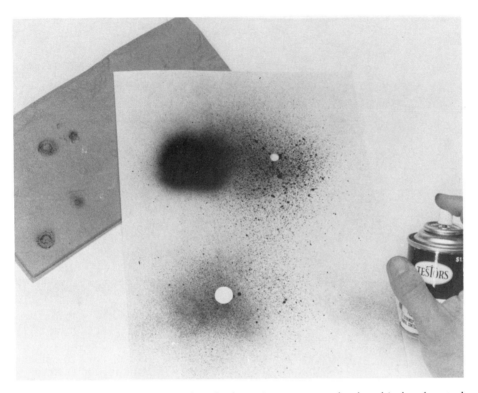

Fig. 3-8 The shadows in craters and other kinds of round shadings can be created with a paper mask with round holes and a quick pass with the spray from an aerosol can of paint.

clearly as highlights. Specific streaks and blotches can then be added as a second or third coat by dabbing on some of the wash of paint and thinner with a sponge, or streaking it on by dragging the sponge over the surface. Figures 3-7 and 3-8 show how holes in a piece of paper can be used to direct the spray for weathering a crater-filled surface and for simulating exhaust or explosion blast stains.

The trick to successful weathering is to use very little paint—use just enough so a viewer will have to look twice to realize that there was any real variation in the basic color of the model. If you apply too much weathering, you'll have to repaint the model, so if you

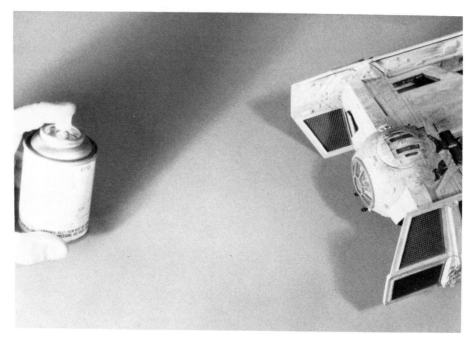

Fig. 3-9 The final finish for any model, after decals and weathering, should be a spray coat of Testors Dullcote or Pactra's Clear Flat.

must err, try to err on the short side. Develop the weathering on a toy or an experimental model until you get the feel of just how much is too much.

Safety

Most of the paints and glues that you will use in model building can be very hazardous either because of flammability, or because of the dangers associated with possible inhalation of toxic vapors. Never use glue or paint near a heater or an open flame. If possible, use a small spray booth. Some model shops carry special cardboard spray booths or you can make your own with a tissue box. Cut one face of the box open and cut a 12-inch square hole in the back. Cover the hole with a furnace filter and mount a common electric fan behind the box so the fan pulls the fumes through the furnace filter. Place the box in front of an open window, or use air conditioner flexible duct pipe to duct it to an open window.

If you're going to do a lot of painting with an aerosol can or an air brush, wear one of the inexpensive mouth and nose covering respirators. Excessive exposure to some of the solvents used in some paints, epoxy, and glues can reportedly cause brain, eye, and other tissue damage. The use of the spray booth, respirator, and the simple expedient of keeping glue and paint bottles covered between each application is very little trouble to go to considering the health problems it can prevent.

CHAPTER 4

Advanced Modeling Techniques

Science fantasy modeling provides a challenge to truly creative modelers to produce a miniature that is something more than just a perfectly finished kit. That challenge can extend anywhere from combining parts of several kits in a process called "conversion" or "kit bashing" to the design and construction of as yet unimagined space vehicles. In every case, the process depends on a strong base of experience in assembling, painting, decaling, and weathering conventional box stock kits.

The materials used for the more advanced models can be either plastic, as in static model rockets and space vehicles, or wood and card, as used for most of the flying model rocket kits. Really it's the end result rather than the materials that matters. There are examples of all three types of construction in this chapter: Jim Mallasch's hand-carved wood, card, and body putty space cruiser, the "Battle Corvette Nebula" of plastic sheet and assorted tank and rocket kit parts, and Robert Angelo's fine Star Wars T.I.E.-Fighter made from parts of the MPC brand Darth Vader℠, T.I.E.-Fighter, plastic sheet, strip, and four plastic pieces from a checker game.

Advanced Modeling Materials

Materials an advanced modeler might use are exactly the same as those that are in your own favorite kits. The difference is that the "advanced" materials don't have the preformed shapes and textures of the kit pieces. Modelers call such pieces "scratch materials." When you scratchbuild, your first goal is to actually make (or scrounge from other kits) the parts that will make a "kit." Once

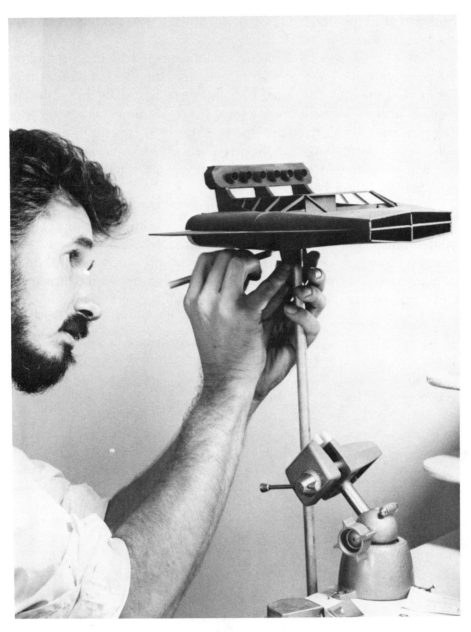

Fig. 4-1 Jim Mallasch prefers aluminum, balsa wood, and card for his commercial rocket models.

you've made your own kit of parts, they are assembled like any commercial kit.

There's a reason professional or advanced modelers, including the pros who make the movie models, use the same materials that are in your favorite kits: they pick the same materials they became familiar with as beginning modelers. There are times when the true professional doesn't have the luxury of his own choice of raw materials, so, in truth, most of the pros are as adept at working with metal as they are with wood and card or with plastic. But each of the pros has his or her favorite medium.

The point I'm trying to make is that you should select materials you are familiar with for your advanced kit conversion or scratch-built models. I made the Battle Corvette Nebula from Evergreen brand .020-inch thick sheet plastic, their plastic strips, and a bunch of plastic tank and rocket parts. If you would rather work with wood and cardboard or, for that matter, with solder-together sheet metal,

Fig. 4-2 This partially completed model by Jim Mallasch, a professional builder, reveals his construction techniques.

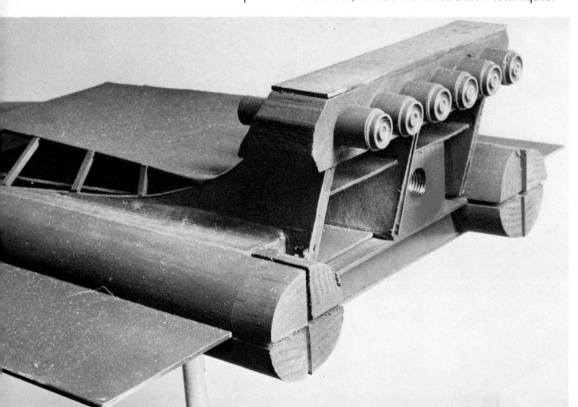

then you should. Jim Mallasch used thick cardboard, balsa wood, and automobile body patching epoxy for his rocket, but you could duplicate it with plastic.

Working with Wood & Card

You would probably develop a liking for wood and card as modeling materials through the assembly of several of the Estes or Centuri brand flying model rocket kits. Similar materials are also used in a number of model railroad car and structure kits and by dozens of flying model aircraft firms. The first model kits in the thirties were wood and card. In fact, until the mid fifties, almost any model kit included only those materials because the current plastic molding techniques didn't exist until that time. Those pioneer modeling materials are still the basis for many of the best model kits on the market, and there are thousands of modelers who prefer them.

The only way you'll learn to work with wood and card is to do it by assembling three or four rockets, flying model aircraft, or model railroad miniatures. If you enjoy all types of miniatures, I can heartily recommend that you try at least one rocket, one flying plane, and one model railroad car or structure kit that is to be made from wood and card. There are slightly different approaches to the use of these materials used by different hobbyists, and you'll definitely learn some trick applications of the pioneer modeling materials.

Modelers who prefer wood and card generally do so because they like the way it feels to cut and shape these materials and because the finished model has a certain weight and mass that is lacking with plastics. In some cases, a professional modeler needs the strength of wood and card construction because the model will be handled frequently. If the model is made for movies or television special effects, wood and card may be needed because the lighting is hot enough to melt the plastics used in most kits.

Flying model rocket kits are made from wood and card because those materials are lighter, in this unique application (particularly the hollow fuselage or rocket body tube), than plastic of similar strength. The injection molded parts used in static model rocket kits are far too strong and far too heavy for good flights, but what is more important is that they can cut right through a person without breaking whereas a balsa or thin vacuum forced piece of plastic would crumble. A WARNING: *Do not ever attempt to use the thick plastic display model kit parts on a flying model rocket!*

There is no advantage whatsoever to limiting yourself to a "pure" wood and card model. That's just one lesson you will learn from building samples of rocket, aircraft, and model railroad kits. The rocket and aircraft models, for example, often use vacuum formed nose cones, canopies, wing tips, and other complex or curved pieces that would be difficult to shape from wood or card. Most of the model railroad kits use cast metal or cast plastic doors that are far more realistic (as well as square and true) than similar items would be if glued together from strips of wood.

Jim Mallasch prefers to use the Bondo-type two-part (filler and catalyst) automobile body repair compounds, like the Easy White brand, for curved areas. He mixes the stuff to a putty-like consistency (it doesn't run like epoxy) and slathers it on the model in piles larger than the final shape. When the Easy White is dry, he can shape and file it like pine without a grain! The entire curved side of his model is solid Easy White.

Jim also prefers the thick pebble grain picture-matting cardboard that is sold by frame shops. Many cardboard enthusiasts prefer the Strathmore brand cardboard because it has a smooth and consistent surface and because it is available in several different thicknesses or plies from .005 to .025 inch. Many wood modelers, including Jim, prefer balsa wood like that sold for building model airplanes. The bass wood from Northeastern, Kappler, and Camino that is sold by model railroad shops is a bit hard to cut, but the bass wood does not have the fuzzy edges of balsa and it's much stronger.

Jim Mallasch, like the professional he is, does not limit himself to just wood, card, and body filler. The window frames on his model are Plastruct brand plastic H-beams and channels, the entire

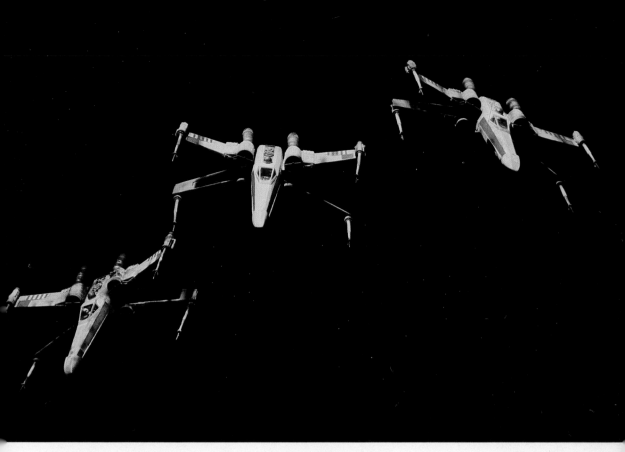

Fig. 1 Robert Angelo's well-weathered Star Wars[TM] X-Wing Fighters[TM] speed through a rhinestone galaxy. A flat black background and low level light were used to eliminate shadows. The models were made from MPC kits, products of Fundimensions, CPG Products Corp.

Fig. 2 The X-Wing Fighter[TM] from the film *Star Wars*. Courtesy of Twentieth Century-Fox. Copyright © 1977 Twentieth Century-Fox Film Corp. All rights reserved.

Fig. 3 The Battle Corvette Nebula is made from plastic "For Sale" signs and tank kit parts. (See chapter 7.)

Fig. 4 The Star Wars™ T.I.E. Fighter™ flying beneath a planet can be simulated easily with MPC's 1/48 scale kit. Robert Angelo built this one. The kit is an MPC Product of Fundimensions, CPG Products Corp.

Fig. 5 The AMT brand kits of the Star Trek™ Klingon Battle Cruiser and the Enterprise were photographed under black light to obtain this deep space effect. Copyright © 1978 Paramount Pictures Corp.

Fig. 6 Robert Angelo added Micro-Scale aircraft decals to his 1/48 scale model of the Star Wars™ X-Wing Fighter™, and weathered the model with thin black paint applied with an air brush. The kit is an MPC Product of Fundimensions, CPG Products Corp

Fig. 7 Model rocketry offers simple-to-assemble scale models and fantasy rockets that are propelled by rocket engines and have automatic parachute or glider recovery systems.

Fig. 8 The Centuri model for the Boeing Air Launched Cruise Missile (ALCM) is an accurate scale model right down to the folding wings, which fold in for rocket flights. All the correct red, white, and blue decals are included.

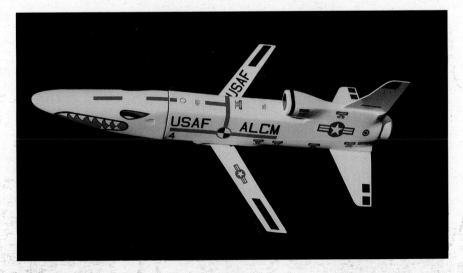

Fig. 9 The Galax IV is Testors' ready-to-fly ground-effect vehicle (hovercraft) powered by an .049-cubic-inch displacement two-stroke model airplane engine.

Fig. 10 The 8-inch-tall Kenner Star Wars™ toy R2-D2™ has a simple two-channel radio for directional movement control. The model is from Kenner Products, CPG Products Corp.

Fig. 11 The Kenner Star Wars™ toy Land Speeder™ performs like a hovercraft or ground-effect vehicle. In the film, the vehicle was floated by an unspecified anti-gravity device. The model is from Kenner Products, CPG Products Corp.

Fig. 12 The effect of real aluminum can be duplicated using Liqu-A-Plate shades of metallic paint. When the paint is dry, it can be polished with a rag wrapped around your fingertip.

Fig. 13 A few small areas of "battle damage" can add realism to a well-weathered fighter. Robert Angelo used an engine from a Lindberg F104 Starfighter for "beneath-the-surface" detail on this Star Wars™ X-Wing Fighter™. The 1/48 scale kit is an MPC Product of Fundimensions, CPG Products Corp.

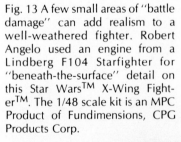

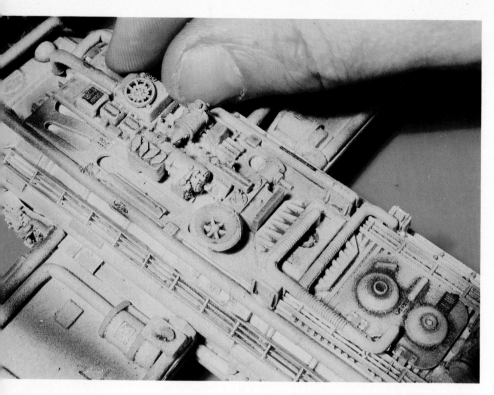

Fig. 14 The complex surface details of the fantasy fighters can be simulated with parts from tank, boat, aircraft, and even HO scale steam locomotive kits. This craft was scratchbuilt by Robert Angelo.

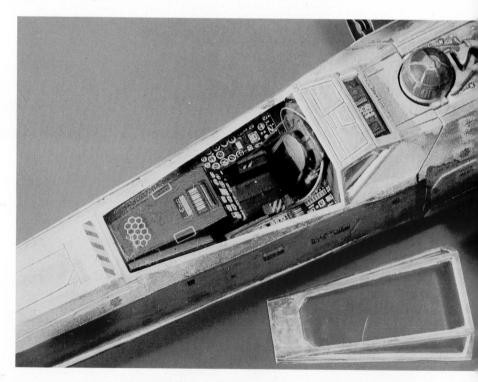

Fig. 15 1/48 scale aircraft decals and pilots can be used to superdetail the MPC Star Wars™ X-Wing Fighters™ the way Robert Angelo did. The kit is an MPC Product of Fundimensions, CPG Products Corp.

wing structure is a sheet of $^1/_{16}$-inch aluminum for strength, and the exhaust tubes are pieces of K & S brand brass tubing (K & S also has aluminum tubing).

The same tools shown in Chapter 2 for working with plastic models will work well for wood and cardboard models. The only extra thing required is several packages of X-acto #11 blades for that #1 knife and several packages of industrial quality single edge razor blades (they're cheaper because they don't have to be sterile). You can cut through a $^1/_{16}$-inch sheet of cardboard in three to six passes *if* you guide the knife or razor blade with a steel straight edge and *if* you keep replacing the blades as they get dull. Mallasch used about 20 #11 blades to cut the parts for that rocket. The choice of X-acto knife or single edge razor blade will depend on your own preferences and on the size of the cut. Balsa wood thicker than about $^1/_8$ inch and bass wood thicker than about $^1/_{16}$ inch should be cut with a razor saw to produce a clean, even edge.

If you expect wood and card to look like metal or space-age plastics of an extraterrestrial vehicle, you'll have to sand all the exposed surfaces and edges with fine sandpaper and apply several coats of model airplane sanding sealer, sanding between each coat. The experience in doing that kind of work on the flying model rocket kits will be a big help here.

Don't be afraid to decorate a display model with parts from plastic kits. The Battle Corvette Nebula and any of its sister ships could just as well have been made from wood and cardboard and detailed with the same tank and rocket pieces as our all-plastic model. Elmer's white glue is the best there is for wood or card model work.

Working with Plastics

Plastic was considered to be junk material suited only for toys until the sixties. Only the true professional modelers were open minded enough to realize that the material could be used anywhere wood or cardboard could and at a considerable savings in time. Plastic cuts more quickly and easily and glues faster than any other material. But it can warp from exposure to heat, and sunlight is its worst enemy even if it's protected by paint. For most display models, however, those disadvantages are not important. Plastic is a bit more expensive than wood, but you won't use enough of it to matter all that much.

Plastic is a lot harder to find than wood. In fact, only the better stocked model railroad and armor model shops carry the Evergreen brand sheet and strip stock or the structural shapes, sheets, and small parts made by Plastruct. You can scrounge much of the plastic you'll need by purchasing plastic "For Rent" or "For Sale" signs (most are .020 to .040 inch sheet stock), using parts of bottles and containers like L'eggs hosiery packages, and parts of plastic model kits, be they for railroads, cars, ships, aircraft, rockets, or fantasy vehicles. Start a plastic scrap box (or a file of boxes for specific

Fig. 4-3 Robert Angelo used a plastic "For Sale" sign, a top from a L'eggs hosiery package, Plastruct ⅛ inch tee plastic, plastic plumbers' washers, railings from Heller's kit for the battleship Tirpitz, plastic checkers, and most of two MPC Star Wars™ T.I.E.-Fighters™ to build his version of the large wing Star Wars™ T.I.E.-Fighter™.

Fig. 4-4 Kenner has a Star Wars™ toy version of this T.I.E.-Fighter™, and Estes has a flying version. Robert Angelo's 1/48 scale model matches the detail on his other MPC brand Star Wars™ kits.

Fig. 4-5 Two MPC Star Wars™ T.I.E.-Fighters™ were used for the basic fuselage of Robert Angelo's model, with a lot of body putty. The centers of each wing are plastic checkers!

shapes and sizes) to hold every leftover part from every kit you buy, including the leftover molding sprues.

Don't be afraid to buy a complete kit just to get, for example, the interior of a wing or landing gear—the remainder can go in the scrap box for likely use on some future modeling project. The weekend swap meets that are often held in drive-in theaters are often sources of older kits and partially completed kits and can supply needed parts at real bargain prices. Try Salvation Army stores and similar shops for kits, too.

You'll soon learn to identify which plastic can be easily cut and glued and which cannot. Most of the true toys are made of a flexible plastic that just won't accept most glues or paints. Even some of the model items like Airfix brand soldiers and other figures are made of this unbreakable, ungluable, and unpaintable plastic. Often, clear plastic sheet and bubbles will be in the ungluable category, but you'll want to use epoxy or cyanoacrylate cement for the clear anyway.

Sheet plastic like Evergreen's and those plastic signs can be cut much like glass. Simply make a light cut along the line you want,

60

then break the plastic over a sharp table edge (or, better, the edge of a 1/4-inch thick Plexiglass work surface) and it will snap along that cut line. You can cut plastic up to 1/8 inch thick with the scribe-and-break method. The broken edge should be filed lightly to remove the rough texture. Even a knife cut or razor saw cut should be filed smooth. Thanks to this quick-cutting technique, you can make parts in minutes from plastic that would take hours in wood or card. This same cutting technique can be used on plastic strips up to about 1/16 inch thick and about 1/4 inch wide. Thicker or wider strip stock should be cut with a razor saw.

Fig. 4-6 The adjustable swivel head of the PanaVise allows the wood support block to be positioned so you can hold the plastic firmly for cutting without damaging the plastic.

I have found that the PanaVise or other brand of swivel head vise is one of the most useful tools made for cutting plastic. Although plastic is too soft to be gripped by even a padded set of vise jaws, you can grip a piece of hardwood at any angle in the jaws, so the wood can serve as a backing for the plastic part. You can then use your fingers as a clamp to hold the plastic to the wood while cutting with a razor saw. Just be sure to use light pressure on the saw and to keep the blade well away from the fingers that are clamping that plastic part. Different thicknesses of wood and different overall sizes can be used to give you a really firm grip on just about any size or shape piece of plastic. The swivel head lets you angle the wood and the plastic to a comfortable position for both the gripping hand and the sawing hand.

Advanced modelers ready to design and make their own may want to turn to Chapter 7 for original designs for the Battle Corvette Nebula and other intraspace vehicles, or to Chapters 6 and 8 for some kit conversion ideas.

CHAPTER 5

Model Rocketry

Model rocketry is both a hobby and a sport. The hobby aspect of the subject involves the basic assembly of rocket models that really can fly when propelled by solid propellant engines. The sport aspect of the subject involves a short trip to the nearest official rocket launching site armed with a couple dozen of the replaceable rocket engines, a launching pad, and some kind of tracking device. On the other hand, a number of modelers would just as soon assemble a flying model rocket simply for its inherent display value.

Flying Model Rockets

The launching of a model rocket is a very accurate reproduction in miniature of what happens at a real-life rocket launch pad. Just like real rockets, model rockets must be trimmed and balanced so they are aerodynamically stable. Testing is a simple process of swinging the model on a string that is described in most kit instructions. A smooth and shiny finish will help the model perform better just as it will the real thing.

The model rocket is propelled by a solidstate rocket engine that is about the size of a tube of lipstick but made of cardboard and ceramic. The design of the engine is what makes model rocketry a safe hobby. The only way to ignite the engine accidentally is to leave it in a fire. A special fuse or "igniter" must be inserted into a cavity in the ceramic end of the engine for normal ignition. The wires from the igniter must then be connected to a launch button which, in turn, is connected to a 6 or 12-volt battery. The electrical current heats the igniter enough so it will fire the rocket engine. A complete new engine is installed for *each* flight.

Fig. 5-1 Clyde Howard's scale model of the Canadian Black Brant sounding rocket leaves the launch pad at the 1978 National Association of Rocketry (NARAM) national finals.

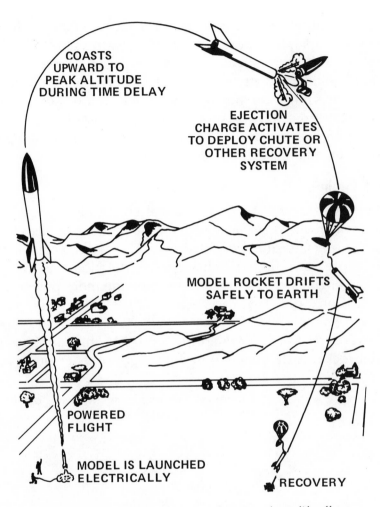

COASTS UPWARD TO PEAK ALTITUDE DURING TIME DELAY

EJECTION CHARGE ACTIVATES TO DEPLOY CHUTE OR OTHER RECOVERY SYSTEM

MODEL ROCKET DRIFTS SAFELY TO EARTH

POWERED FLIGHT

MODEL IS LAUNCHED ELECTRICALLY

RECOVERY

Fig. 5-2 The typical sequence of events, from lift-off to recovery, in the launching of a flying model rocket. *Courtesy Estes Industries.*

A guide rod on the rocket launch pad directs the rocket on a straight upward path for that critical first few feet and then the rocket blazes its own trail. After a predetermined number of seconds, the rocket engine finishes its "burn" and allows the rocket to coast for a moment before a reverse charge in the engine forces the nose cone from the rocket to release a parachute that gently lowers the rocket, engine, and nose cone back to earth. Alternate recovery systems include long streamers and gliders or a combination of parachute, glider, and streamer. Some rockets release one or more gliders at the same time they deploy the parachute; others are two-stage rockets that have a second or third rocket engine that is started when the first stage is completed. It's that kind of flight and action variety that makes this such a popular hobby.

Rocket Engines

There are dozens of different rocket engines in four basic sizes. Each size of engine is available with a wide range of choices in terms of both power and duration of thrust. The larger engines are, of course, more powerful and/or have longer duration than the smaller ones. The engines can be clustered to provide more power or used with a two or three stage rocket design. The latter is usually more effective.

The Federal Aviation Administration (FAA) limited the weight of a model rocket to 16 ounces and the weight of its engine's propel-

Fig. 5-3 A group of photographers witnesses one of the lift offs at the 1978 NARAM national meet.

lant to 4 ounces. An engine with a fourth of that propellant is enough to get a well-designed rocket over 2000 feet. The largest rocket models go out of sight of the naked eye at about 2000 feet and most models disappear at 1000 feet or more. In other words, the available engines are quite large enough.

Most states, counties, and municipalities have laws that regulate where you can fly a model rocket, and many of them require that you have a permit from the forest service, fire department, or some other authority. The permit generally is intended to see that an adult is present when any youngster flies his rockets and that you know just where and when it is legal to fly. A local hobby dealer who sells model rocket kits should also be able to tell you where to fly. If not, contact either Estes or Centuri directly.

Rocketry Safety Code

The National Association of Rocketry, the National Fire Protection Association, The Hobby Industry Association of America, and the model rocket manufacturers themselves have developed and endorsed the following Model Rocketry Safety Code:

1. *Construction*—My model rockets will be made of lightweight materials such as paper, wood, plastic and rubber, without any metal as structural parts.
2. *Engines*—I will use only pre-loaded factory made model rocket engines in the manner recommended by the manufacturer. I will not change in any way nor attempt to reload these engines.
3. *Recovery*—I will always use a recovery system in my model rockets that will return them safely to the ground so that they may be flown again.
4. *Weight Limits*—My model rocket will weigh no more than 453 grams (16 ozs.) at liftoff, and the engines will contain no more than 113 grams (4 ozs.) of propellant.
5. *Stability*—I will check the stability of my model rockets before

their first flight, except when launching models of already proven stability.

6. *Launching System*—The system I use to launch my model rockets must be remotely controlled and electrically operated, and will contain a switch that will return to "off" when released. I will remain at least 10 feet away from any rocket that is being launched.

7. *Launch Safety*—I will not let anyone approach a model rocket on a launcher until I have made sure that either the safety interlock key has been removed or the battery has been disconnected from my launcher.

8. *Flying Conditions*—I will not launch my model rocket in high winds, near buildings, power lines, tall trees, low flying aircraft, or under any conditions which might be dangerous to people or property.

9. *Launch Area*—My model rockets will always be launched from a cleared area, free of any easy to burn materials, and I will only use non-flammable recovery wadding in my rockets.

10. *Jet Deflector*—My launcher will have a jet deflector device to prevent the engine exhaust from hitting the ground directly.

11. *Launch Rod*—To prevent accidental eye injury I will always place the launcher so the end of the rod is above eye level or cap the end of the rod with my hand when approaching it. I will never place my head or body over the launching rod. When my launcher is not in use I will always store it so that the launch rod is **not** in an upright position.

12. *Power Lines*—I will never attempt to recover my rocket from a power line or other dangerous places.

13. *Launch Targets & Angle*—I will not launch rockets so their flight path will carry them against targets on the ground, and will never use an explosive warhead nor a payload that is intended to be flammable. My launching device will always be pointed within 30 degrees of vertical.

14. *Pre-Launch Test*—When conducting research activities with unproven designs or methods, I will, when possible, determine

their reliability through pre-launch tests. I will conduct launchings of unproven designs in complete isolation from persons not participating in the actual launching.

Launching Systems

A scale model rocket can be launched using a small piece of sipping straw on the side and a launching guide rod. Many modelers, however, prefer to keep the exterior of their rockets as clean as possible for both appearance and the slightly improved performance that results from removing the friction-producing tube.

Fig. 5-4 The wire hook on this launching lug is a slip fit into a notch in the side of the rocket body. The lug stays on the launch tube, not the rocket, at lift-off.

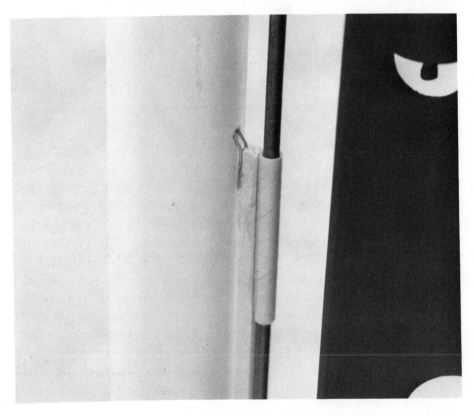

Fig. 5-5 Brass and aluminum pins and brads simulate rivets on many of the scratchbuilt flying scale rockets.

Two solutions are currently in use by NAR members: a small wire mounted slip lug that is essentially a piece of sipping straw glued to a strip of wood with a small hairpin-size loop of wire epoxied to the end. The wire loop engages a small slit in the side of the rocket. A rubber band at the top of the launching rod stops the sipping straw as the rocket is launched and the slip lug disengages from the rocket's notch so the rocket can continue on its journey upward. The wire must be a clean fit on the rocket body to hold it firmly for the initial few feet of acceleration but loose enough to disengage cleanly without pulling the rocket off course at the top of the launching rod.

An alternate method of launching scale model rockets is the homemade tri-bar launching stand (figures 5-7 and 5-8). The top and

Fig. 5-6 You can barely see the clear plastic fins needed to add stability to a scale model of the Saturn V.

bottom are ¼ x 2 inch rings about 10 inches in diameter. Aluminum is best, but steel could be substituted. There's a good chance you could find a light fixture in a lighting supply shop that would be close enough to the right size that you could make the top and base from it. The outer vertical members are ½ inch aluminum angles, and the three center bars are approximately ⅜ inch aluminum tubes.

The tubes are held in place with bent pieces of ⅛ inch threaded

Fig. 5-7 This three-foot-long launching platform is designed to guide scale model rockets at lift off without the need for out-of-scale launching lugs on the rocket bodies.

Fig. 5-8 The bolts and wing nuts at the base of the tri-bar launching pad allow it to be adjusted to fit any diameter rocket body that has three fins.

rod inserted into tight-fitting threaded plugs in the aluminum tube ends. Coil springs hold the three tubes toward the center of the launching fixture. Wing nuts on the outside keep the coil springs in tension. The three aluminum tubes can be moved farther away from or closer to the center by simply loosening or tightening the wing nuts on the outside. The nuts visible on the inside of the large rings are just spacers that slide over the threaded rods. The wing nuts are adjusted at both top and bottom so the three aluminum tubes just clear the body of the rocket model. The spaces between the tubes and the large inside diameter of the top and bottom rings (about 10 inches) provide ample clearance for the three-fin placement common to most scale model rockets.

The whole launching rig is mounted on a camera tripod so it can be leaned into the wind as needed. The three foot aluminum tubes provide ample guidance for the rocket's initial acceleration and no launch lugs or slots are needed on the rocket itself. The aluminum parts, threaded rods, nuts, springs, and other hardware should be available at any hardware store.

As you might guess, scale model launching systems are available to use with scale model rockets. Numerous local and regional rocketry contests and an annual National Association of Rocketry contest called the NARAM include categories for both scale model rockets and for scale launch systems. Models of the Nike and Tomahawk launching systems have been popular. The complex-appearing tower with crescent-shaped counterweights (figure 5-9) is a scale model of the Australian Aeolus launcher and rocket built by the team of Pearson, Steele, and Nowak for the 1978 NARAM event.

Tracking Rocket Flights

You can use a simple protractor to determine the altitude of your rocket's flights. Mount the protractor through a hole drilled in the center of the straight side to a board that can be clamped to a camera tripod or similar stand. Sight along the straight edge of the protractor at the rocket and swing the protractor up around its pivot hole to follow the rocket's path upward with your eye still sighting down the protractor. Hold the protractor in position when the rocket has reached its apogee and you will have the angle between the rocket's peak and the place where you placed the protractor-tracking device. Hang a plumb line (a weight on a string) from the protractor's pivot point to read the correct number of degrees of the angle.

Knowing the angle of flight and the exact distance from the launching pad to the protractor, you can then determine the rocket's height using simple geometry. The rocket's vertical path forms a 90° angle with the ground. The protractor gives you a sec-

Fig. 5-9 Scale model launching pads bring an aura of realism right to the moment of lift off.

ond angle. With that 90° angle of launch, the protractor's measured angle (A), and the distance to the launch pad (D), figure the height (H) of the vertical leg of the triangle (the rocket's height) with the following formula: H = D × tangent of angle A. You will need a table of tangents from any high school or college algebra, geometry, or trigonometry textbook to complete the equation.

There will be some error in the measurement because wind and drift will force the rocket to take something less than a 90° path upward from the earth. You can minimize that error by placing the protractor tracking station at least 100 feet from the launching pad and exactly 90° to the path of the wind. Any wind gusts will divert the rocket from a true vertical course, but your measurements will be about 90 percent accurate as long as the rocket is not being blown any closer to or further away from the tracking station. If the wind direction shifts, move the tracking station to maintain that 90° angle to the wind. Be sure to keep the station the same distance

Fig. 5-10 This handmade plywood box has special foam-padded partitions to transport flying scale model rockets without damage to the rockets themselves. The engines are never installed until launch time.

Fig. 5-11 This fine theodolite can be constructed with three plastic protractors and some scrap materials. It will help to provide accurate rocket tracking information.

from the launching pad each time you move it so your geometric calculations will be accurate.

You can eliminate most of the error that is created when the rocket's course is diverted by the wind if you set up two tracking stations and utilize slightly more complicated tracking devices called theodolites. Theodolites are virtually identical in function to the sextants that sailors use for celestial navigation. Each tracking device (theodolite) will measure the horizontal angle between the tracking station and the rocket's apogee as well as the vertical angle. Each tracking device needs to be fitted with two more protractors, mounted base-to-base, to form a circle (see figure 5-11). If you sight along the altitude tracking tube you can also determine the horizontal angle by aligning the marks on the horizontal pair of protractors. You can almost determine with just a glance at the photograph how to make this tracking device. The three protractors are held in place with 2 x 2 lumber scraps and wood screws. Mailing tubes and wire

Fig. 5-12 Two trackers with theodolites are stationed at each of two tracking stations at all NAR national contests to insure reliable information on each rocket's flight.

Fig. 5-13 The cross-hair sight on this theodolite makes it a bit easier for the tracker to keep the rocket in view to the apogee of its flight.

form a cross-hair sight on an 18-inch length of 2 x 2 to make sighting the rocket during its flight easier. The double layer of Masonite with screws, spacer springs, and wing nuts (figure 5-11) is an optional method of providing precise leveling. Adjusting the camera tripod's legs individually would have the same effect. The screw through the curved part of the protractor is there to hold the protractor to the vertical support. The cross-hair sight and its 2 x 2 board are pivoted upward to follow the rocket's path off the launching pad. A piece of wire on the 2 x 2 points to the proper degree of vertical angle on the vertical protractor. It is important that the theodolite be adjusted so that the wire pointer reads 0° when the cross-hair sight inside the mailing tube is aligned with the rocket's nose on the launching pad.

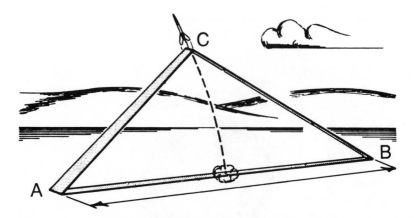

Fig. 5-14 Using two tracking stations (A and B) located directly in the path of the wind and exactly 1000 feet apart will improve the accuracy of your calculations. *Courtesy Estes Industries.*

The two tracking stations must be positioned directly in the path of the wind rather than at 90° to it. Place station A exactly 500 feet from the launching pad and station B 500 feet on the opposite side of the launching pad. You can use the horizontal protractors on each theodolite to help align them along that imaginary straight line. The altitude of the rocket is determined utilizing (1) angles A and B as measured by the theodolites at tracking stations A and B, (2) angle C, which is simply the sum of angles A and B deducted from 180°, (3) the distance between the stations, which you know to be 500 feet plus 500 feet, or 1000 feet, and (4) a geometry book to find the sines of angles A, B, and C. The following formula will then give you the height of the rocket's flight:

$$\frac{distance\ between\ stations \times \sin A \times \sin B}{\sin C} = height\ of\ flight$$

You may need to shift the location of both tracking stations each time the wind changes to keep the stations directly in line with the path of the wind and the launch pad.

Rocket Kits & Interspace Vehicles

There are two types of model rocket kits: those that are freelance designs just for superb flights and those that are models of real and imaginary (fantasy) space vehicles. In most cases, the scale models and fantasy fliers are much more fascinating to look at than they are to fly. There is, for instance, a scale model kit for Star Wars R2-D2 by Estes that makes a nice-looking nine-inch-tall miniature. It was never designed to be aerodynamic, though, so its flights are of very short duration. Large clear plastic fins must be added to make it stable, and it has the unnerving habit of popping its rounded top to deploy the parachute at the peak (apogee) of its upward flight. The Estes model of the Canadian sounding rocket Black Brant III, by way of contrast, is one of the best-flying rocket kits available. In general, though, you'll have to choose between the super quick, ultra high flights of the freelance competition model rockets and the somewhat more sluggish performance of way-out fantasy designs or super scale models.

I happen to prefer the fantasy designs because I like the rockets I make to fly slowly enough and to reach apogee at a low enough altitude so I can get a nice long look at the results of my modeling efforts in action. I mentioned earlier that model building is a thrilling, three dimensional art form. The additional aspect of authentic action makes model building the most fascinating hobby possible.

Basic Kits

The various flying model rocket kits available span a complete range from those that challenge the skill of an expert down to snap-together models that can be ready to launch in as little as 15 minutes. The quickies are pre-painted so that all you have to do is snap the parts together or at most glue the fins or rocket holder in place. Most of the kits even feature pre-cut balsa wood fins. The modeler is required to sand the fins to an airfoil or teardrop shape (as viewed from the outer edge of the fin), to glue the fins to the cardboard body, and to paint the model with sanding sealer and color coats, sanding after each coat except the last to get a super smooth finish. The various tabs and brackets (usually cardboard) that hold the rocket engines must also be glued inside the body.

The Centuri and the Estes rocket catalogs rank their kits according to the skill required, ranging from level 1 for the simple rocket

Fig. 6-1 The U.S. Air Force F-12 interceptor and the identical-looking SR-71 reconnaissance jets are over 20 years old, but they still look like something yet to come. *U.S. Air Force photo.*

Fig. 6-2 Flying model rockets range in size from pencil-size mites to almost life-size models like Estes' Mean Machine. *Courtesy Estes Industries.*

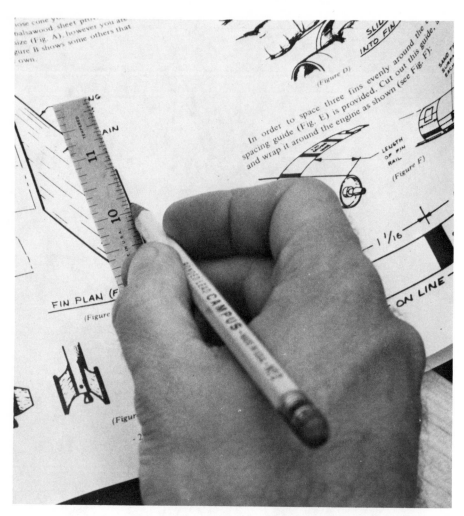

Fig. 6-3 Most kits have pre-cut fins, but some require you to trace through a paper pattern with a pencil to mark the fins on a sheet of balsa. Use a sharp hobby knife to cut the balsa.

kits to level 5 models like the Saturn V or Mercury Redstone. Some of the detail parts on the complex-appearing kits are plastic. The size of the finished models ranges anywhere from a 7-inch Estes' Scout to a relatively easy to build (skill level 3) 1/19 scale model of the German V-2 that is almost 3 feet tall and 4 inches in diameter.

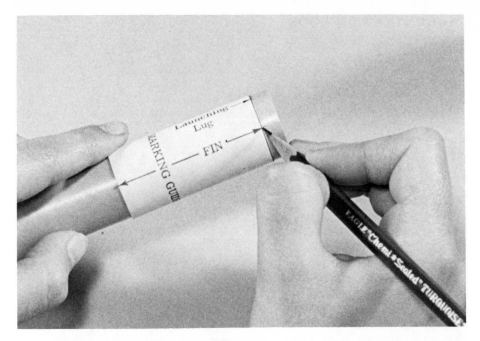

Fig. 6-4 Most kits include paper patterns to assure that the fins are spaced equally around the cardboard body tube. *Courtesy Estes Industries.*

Fig. 6-5 Sand each of the fins to a perfect shape and taper the ends, then glue them to the body with white glue. *Courtesy Estes Industries.*

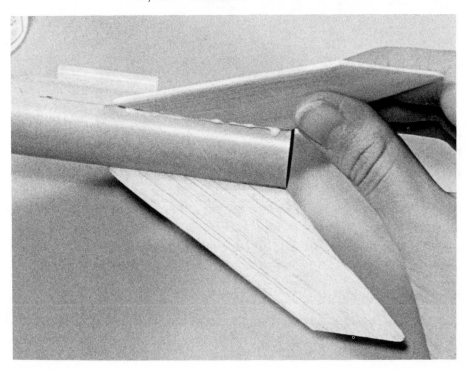

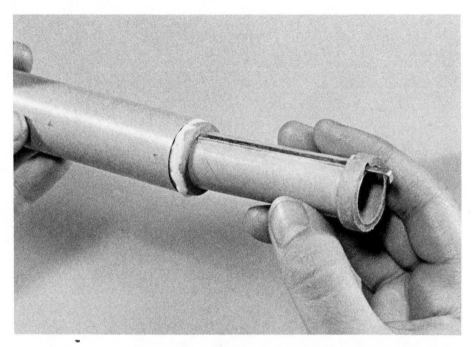

Fig. 6-6 A second cardboard tube with some spacers and a metal clip is assembled and glued inside the rocket body as a support for the rocket engine. *Courtesy Estes Industries.*

If you expect to build a reasonable replica of the Star Trek Klingon Battle Cruiser or Starship Enterprise, or the Star Wars T.I.E.-Fighter, X-Wing Fighter, or R2-D2, then you'd best hone your skills because those Estes models are skill levels 2, 3, and 4. The complex Estes, Centuri, and Flight Systems rockets do provide ample rewards for the efforts expended in completing them. They really provide more of a feeling of pride and accomplishment than you get from assembling an all-plastic kit, and it really doesn't take much longer to complete one of the flying rockets.

Interspace Rockets

The fascination that man has for rockets began, of course, with imaginary flights from Earth to somewhere out there in space. *In-*

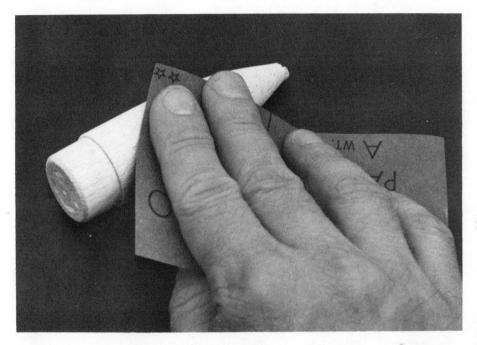

Fig. 6-7 Use fine sandpaper to smooth the fins and the nose cones and finish them with several coats of sanding sealer, followed by several coats of paint, sanding between each coat.

terspace rockets were the vehicles for that type of Earth-to-space travel. The Buck Rogers type of rocket also fitted that mold. *Intraspace* rockets, featured in the next chapter, are the product of a later time when we earthlings began considering space vehicles whose flights started and ended in the weightlessness of space. The difference to the modeler lies in the appearance of the rocket. The interspace rockets must have some hint of aerodynamics because they must fight their way through Earth's (or some other planet's) atmosphere to reach space. The term "rocket" conjurs up something like the German V-2, the George Pal *Destination Moon* rocket, or the Saturn V in most of our minds. Vehicles like those in "Star Trek," *2001: A Space Odyssey,* and *Star Wars* are generally thought of more as space vehicles rather than old-fashioned rockets. The

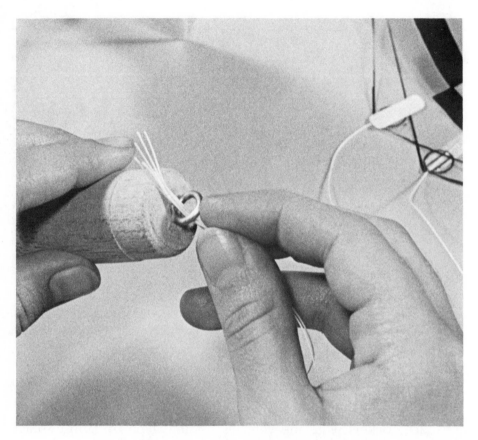

Fig. 6-8 A rubber band "shock cord" connects the nose cone to the rocket body. A screw eye at the base of the nose cone is used to anchor the recovery parachute and the shock cord. *Courtesy Estes Industries.*

docking maneuvers in space and the unforseen characteristics of the vehicle used to land the first man on the moon have changed our thoughts about true space vehicles.

It might be expected that the firms that produce the flying model rocket kits would have been way ahead in designs for interspace rockets that must contend with the earth's atmosphere since that is one of the problems their standard products face all the time. Surprisingly, though, neither Estes, Centuri, Flight Systems,

Fig. 6-9 There are a half dozen display models of the Space Shuttle, but this one, from Estes, really does fly *and*, at apogee, the rocket portion returns via parachute while the Shuttle glides down. *Courtesy Estes Industries.*

Fig. 6-10 The Estes Star Wars™ X-Wing Fighter™ Starter Kit includes a wood, card, and plastic flying model (shown both as the kit *and* assembled), launch pad, and three rocket engines (lower left). *Courtesy Estes Industries.*

nor any of the firms that used to compete with them came very close to the designs the film magicians created for *Star Wars* or "Space: 1999." When models of these fantasy vehicles were produced as prototypes for the kits of these vehicles that really would fly, the kit makers found out why the film makers' designs looked so revolutionary—the fantasy vehicles really don't fly worth a darn. The Star Wars X-Wing Fighter may look like a much modified jet of the future, particularly with that open wing capability for in-space maneuvers, but it doesn't fly with much stability as a rocket without some additional clear plastic trim tabs.

The flying model kit makers did predict the shape of the real Space Shuttle to an amazing degree. Estes' scale model of the actual Space Shuttle is visible as a stubby version of several similar glider and rocket models in their catalog and similar "birds" in Centuri's catalog. The difference between the current trend of motion picture special effects and reality like the Space Shuttle is aerodynamics. If a vehicle operates in a gravity-free environment as well as under conditions of virtually no atmosphere, there is very nearly nothing to retard its acceleration or its flight except its own weight and mass. A square block would fly as efficiently from the moon to the sun as the Space Shuttle or any other aircraft-style, streamlined craft.

Only the X-Wing Fighters in *Star Wars* make any pretense of being flyable in atmospheric conditions, thanks to the wing splitting design. The scenario for the film required them to take off and land on a planet with atmosphere, yet still be capable of efficient maneuvers in space. There was no visible attempt made to make the T.I.E.-Fighters aerodynamic. Frankly, the vehicles could have been square blocks, and the fine filming and special effects work would have made them appear to be not only feasible but immediately real. The fine model work completed the illusion.

The amateur model maker won't have the advantage of the world's finest film technicians in making his models realistic through simulated movement. If you're making a model of one of the television or motion picture models, then you'll need to go well beyond the unpainted plastic to provide weathering and shading ef-

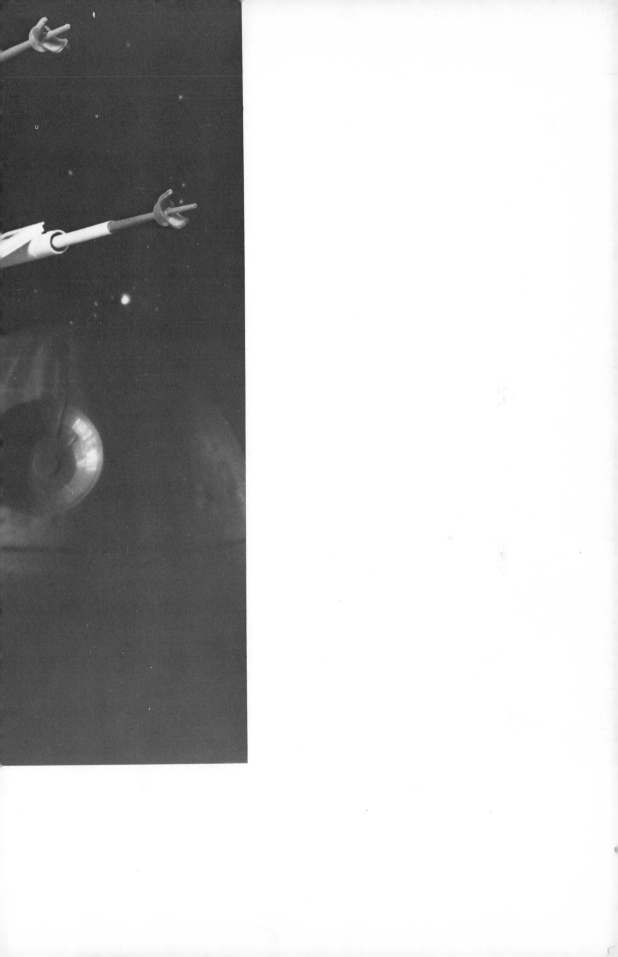

fects and perhaps even special on-vehicle or display lighting to come close enough to the feeling of the real thing. A particularly fine diorama or some special display illumination can go a long way toward matching the images you and your friends recall from the motion picture and television scenes. There are some ideas on that subject in Chapter 11. If you are going to design your own space vehicles then you'll have to consider credibility even more than the movie model makers because you don't have a special effects crew to back you up.

Designing Your Own

You must decide on a flying rocket *before* you built it. Do not attempt to make a flying rocket out of one of the model kits or the toy display rockets, because it just won't fly. Those all plastic toys or models might, if forced to fly, explode like a bomb. The hobby of flying model rockets is as well established as the plastic kit hobby and, if you follow the safety code and stick to the basic design principles of the hobby, flying model rockets is as safe a hobby as there is.

Your best bet for an original model design is to work with one of the flying model rocket fantasy designs or to take one of the historical rockets and modernize it with some laser style guns. A few examples of flying rocket designs are shown in these pages and more are in the Estes, Centuri, and Flight Systems catalogs. It's far easier to begin with one of the flying rockets for your own design because the major shaping is already done for you and, of more importance, the design has to be credible because it really was meant to fly!

The rockets that can be considered historical include those used in motion pictures and television. Photos of those are available in books at your local library as well as in the various magazines dealing with Sci-Fi films. You should find some inspiration, at least, from the various films George Pal produced in the fifties, and you may even like the designs used for TV's "Lost in Space" or "Land of the Giants."

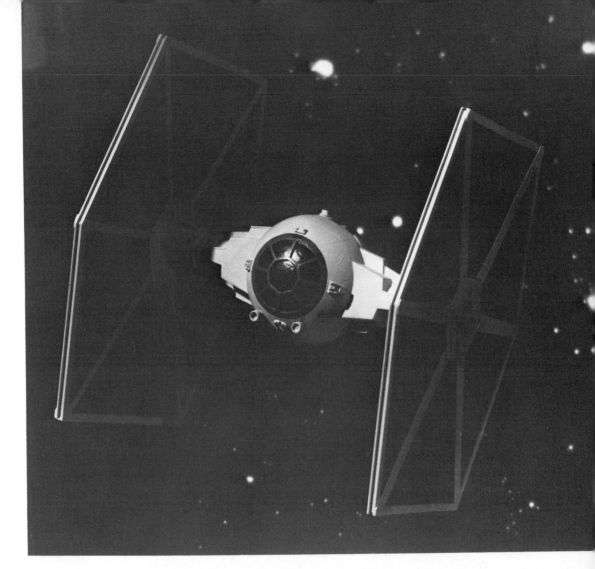

Fig. 6-12 The Star Wars™ T.I.E.-Fighter™ kit, from Estes, is a plastic, wood, and card kit to build a rocket-powered version of the motion picture vehicle. *Courtesy Estes Industries.*

There were kits for some of these older designs but they're now expensive collector's items worth $10 or more as kits and worth next to nothing if assembled. If you do decide on one of the older fantasy designs you'll probably have to scratchbuild it.

Historical Rockets

Historical rockets are often more fascinating than someone's fantasy design. Some of the first flying rockets of the thirties and forties looked very much like the designs that DaVinci and the Buck Rogers creators had predicted. Fortunately for the rocket modeler, the firms that make plastic display model kits have included most of the historical rockets in their lines of historical aircraft. You will definitely have to search for the models because some are on and off the market several times during a five year period, some are imports only available at hobby shops that specialize in armor and aircraft, and others are marketed mostly through five-and-dime stores and drugstores.

The famous German V-2 rocket has been offered, complete with interior detail and a transporter, by Revell in 1/69 scale, Eidai made a 1/76 scale version that included both the transporter and a truck, and Faller makes a 1/100 scale kit that includes a model of the V-1 rocket as well. Heller has a 1/72 scale Fieseler F1 103 that looks like the double-pod V-1 with a pilot's canopy and a Bachem Ba 349 A

Fig. 6-13 The Kenner Star Wars™ toy version of the T.I.E.-Fighter™ has slightly smaller solar panel wings than it should to be an accurate scale model, but it's a good start toward one.

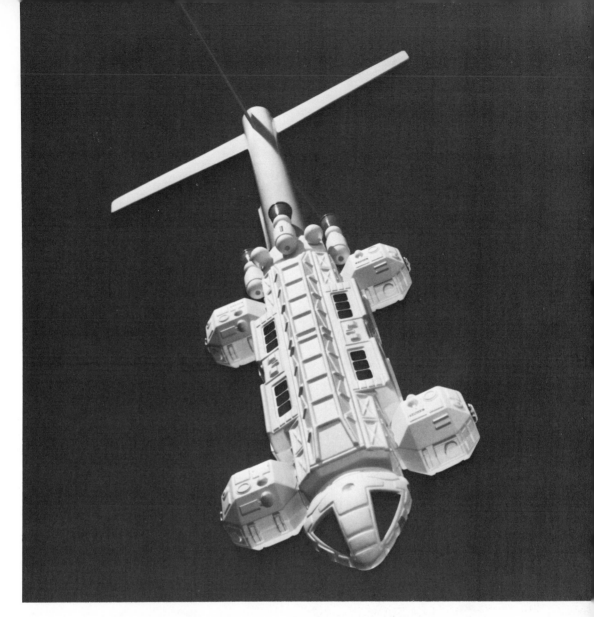

Fig. 6-14 The Eagle from "Space: 1999" really flies, with the help of a tail extension and some fins, when you complete Centuri's rocket model kit of the vehicle. *Courtesy Centuri Engineering.*

"Natter" that looks like a rocket version of the famous Me109 Messerschmitt fighter.

The Heinkel He162 WWII jet is available in 1/72 scale from Lindberg and 1/100 scale from Faller. A gigantic (56 inch wingspan) balsa wood kit for a flying model of the He162 is offered by Midwest along with a ducted-fan engine to match. The bat-like little Messerschmitt Me163 Komet was available in 1/48 scale from Hawk and is still offered in 1/72 scale by Lindberg. Hawk once had a 1/41 scale

Fig. 6-15 Even the Star Wars™ R2-D2™ android can be assembled into a flying model rocket with the help of this Centuri model kit. *Courtesy Centuri Engineering.*

Fig. 6-16 Jet-powered fighters have been redesigned for rocket power in the Fighter Fleet series from Centuri. *Courtesy Centuri Engineering.*

model of the strange Japanese bomber-launched Kugisho MXY-8 Baka jet, and the similar Ohka rocket is included in the Mini-craft/Hasegawa 1/72 scale Mitsubishi Betty bomber kit.

America has some strange early rockets of her own, including the Bell X-1B. Stombecker had a 1/48 scale kit, but it's so rare you'd be better off to carve your own from balsa. The dual-engined Lock-heed YF-12A interceptor is one of the best possible sources for parts for new and different do-it-yourself rocket designs even though this one was really a jet powered craft. Revell offered a finely detailed model of this aircraft as the Lockheed SR-71 reconnaisance version. The first true interspace rocket, the North American X-15, is available as a 1/68 scale flying model rocket (but in plastic!) from Estes and as a display model kit in 1/65 scale from Revell and 1/130 scale from Heller.

If you want to prepare your own interspace rocket design kit, I would suggest purchasing as many of these German, Japanese, and American prototype early rocket models as you can. It would be simple enough to use the plastic modeling techniques in this book to combine parts from two or more of these kits to make a rocket that would rival the best the motion picture studios have been able to devise. Your model, however, will have that hard-to-define look of airworthiness because it will be based on rocket designs that re-ally did fly.

Most of the current rockets that the armed forces consider to be war machines are pretty mundane looking, but there are some notable exceptions like the 1/24 scale Estes flying version of the Air Force BOMARC rocket. Centuri has modified many of the current jet fighters (in just the way they would have to be changed in real life) for operation and stability as flying rockets. You might learn something useful from their Fighter Fleet series that includes the F-15, F-4, and F-16 adapted to rocket design principles.

Scale Models

Fortunately for budding rocket designers, the line of Estes and Centuri kits includes just about every desirable scale model of a real space vehicle as well as models of the popular fantasy vehicles.

Fig. 6-17 The 5200 Rocketry Exploration set from Centuri includes an array of interchangeable fins, nose cones, payloads, and engines. *Courtesy Centuri Engineering.*

Estes has kits for the Star Wars X-Wing Fighter in two sizes (9- or 16-inch wing span) as well as the T.I.E.-Fighter and R2-D2, the Star Trek Klingon Battle Cruiser and Starship Enterprise, and Centuri has the Space: 1999 Eagle.

Estes has flying scale model kits for the new Space Shuttle, the

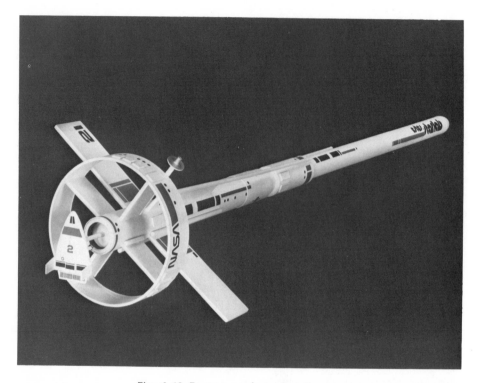

Fig. 6-18 Fantasy rockets that fly are a specialty with firms like Estes. This is their Star Lab. *Courtesy Estes Industries.*

World War II German V-2, the X-15, Little John, Pegasus, BOMARC, the Russian Wolverine, Bandit, Nike-X, WAC Corporal, Canadian Black Brant III, Nike Ajax, Saturn V, LTV Scout, Honest John, Mercury Redstone, Pershing 1A, Saturn V, and the Mars Lander craft.

Centuri's list of scale models includes the Israeli Gabriel, Italian Sea Killer, Russian SAM-3, and America's Nomad, Jayhawk, Scram-Jet, Nike Smoke, Mercury Redstone, Saturn 1B, Saturn V, Cruise Missile, and an array of jetlike rockets like the five shown in figure 6-16. Centuri also has a Rocketry Exploration outfit that includes two rockets with interchangeable fins, nose cones, payload sections, and engines, with a launchpad and an assortment of engines. The instruction booklet in the Centuri exploration outfit offers a short

Fig. 6-19 The flight of fantasy is apparent in Estes' Satellite Interceptor flying model rocket. *Courtesy Estes Industries.*

course in what works and what doesn't in conventional rocket designs.

Centuri also has a unique array of special rocket and rocket-glider designs like the X-24 Bug, Mach 10, Fireflash, Vulcan, E.S.S. Raven, and the U.F.O. Invader. Estes has the Mars Snooper II, Orbital Transport, Scissors-Wing Transport, Constellation, Starlab and Solar Sailor.

I suggest you amass a sizable amount of experience building flying model rocket kits before deciding to try to design and fly your own. The model makers of the flying rockets—Estes, Centuri, and Flight Systems—have their own design teams plus a few thousand rocket enthusiasts developing and testing various designs. You'll have a tough time coming up with a stable rocket design that is anything but a variation on the already proven Estes and Centuri kits.

Estes and Centuri, offer a collection of booklets that includes just about everything from how to calculate aerodynamic drag ($1.25 from Estes) to altitude performance ($1.00 from Centuri). There are several other publications, in addition to the design and research books offered by Estes and Centuri, that will give you more information on rocket design, rocket flights, launch systems, tracking systems, and recovery systems. The best is the one written by model rocket pioneer G. Harry Stine called the *Handbook of Model Rocketry*. Stine and Orville H. Carlisle invented the model rocket in 1957, Carlisle having flown his first experimental prototype in 1954. The book is $6.95 and available at hobby shops or by mail from Estes or Centuri. The National Association of Rocketry publishes the monthly *Model Rocketeer* for $7.00 a year, or you can get the magazine as part of your membership in NAR: Those under 16 pay $7.00 a year dues, those under 21 pay $8.00, and those over 21 pay $10.00 to P.O. Box 725, New Providence, N.J. 07974. *Model Rocketeer* also frequently publishes plans of unusual rockets and launching systems.

Flying Saucers

The ancient concept of vehicles visiting from outer space as saucer-shaped craft persists, heightened, perhaps, by the success of

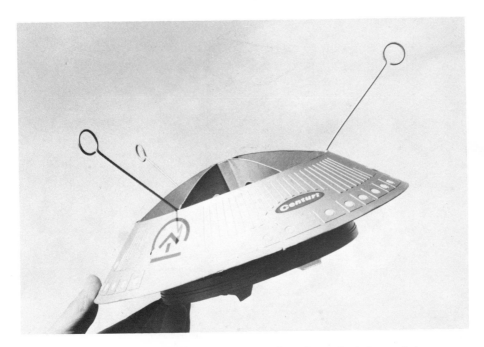

Fig. 6-20 This wood, card, and metal miniature flying saucer is an easy-to-build model that really does rocket straight up for a gliding, saucer-like recovery.

the film *Close Encounters of the Third Kind*. The vehicles from that film were more platforms for lights than vehicles, but even that effect can be duplicated with the special lighting effects described in Chapter 11.

There are at least two models of flying saucers that fly, each, astoundingly enough, on totally different principles. Cox has a 12-inch-diameter, ready-to-fly saucer called Star Cruiser-UFO that has a fuel-powered .049-cubic-inch displacement engine and propeller. The free flight model combines the flight principles of hovercraft, ducted fan jets, helicopters, and airfoils for its flights. Centuri's Flying Saucer is a card and wire kit that is lifted vertically by rocket power for a glider-like flight back to earth. There have been several saucer-like vehicle kits from older television shows, but the only kit currently on the market for a plastic display model of a flying saucer is Lindberg's number 1152 UFO kit.

Fig. 6-21 The Cox Star Cruiser-UFO looks like a flying saucer and flies on the principles of a helicopter thanks to a fuel-burning model airplane engine. *Courtesy Cox Hobbies.*

The varieties of spun aluminum cooking and serving ware are probably a better source of flying saucer shapes for the custom designer than plastic plates or dishes. Just about any of the flying or static rocket kits would provide detail parts for a scratchbuilt flying saucer miniature.

CHAPTER 7

Intraspace Vehicles

"Like the product of an imagination run wild" is one way to describe any vehicle designed to fly in the physical freedom of space. The *intraspace* vehicles, by definition, are transports designed to begin and end their flight beyond the confines of any planet's atmosphere or gravity. It's a concept that is very close to the limits of man's imagination because the environment of space is something we have only just begun to experience.

The first intraspace-like vehicles were probably the lunar modules like the Eagle and even it had to contend with the gravity of the moon. Those lunar modules held the visual excitement of the Wright Brothers' Kitty Hawk flier (and well they should) for either type of vehicle represents nothing less than the threshhold of a whole new concept of travel. The motion picture *2001: A Space Odyssey* probably came as close as man could to assigning dimensions and designs to intraspace vehicles of almost every type from one man modules to city size transports. Stanley Kubrick's people carried imaginary intraspacial flight almost to reality with their three-dimensional miniatures and special effects. The ultimate development of the lunar Eagle school of spaceship design is, for the present at least, the machinery in *2001*.

Designing Custom Spaceships

The only constraints that face the designer of an intraspace craft are the need for life support systems for the inhabitants and some form of action-reaction propulsion that includes, but is certainly not limited to, rocket engines. There is no need for the conventional

Fig. 7-1 The Star Trek™ Enterprise is one of the all-time classic intraspace vehicles. This is the AMT kit for the vehicle. *Courtesy AMT Corp.*

rocket's pointed nose or teardrop profile because there is no sea of atmosphere to slice through. There is no need for the smooth exterior surfaces that man has had to work very hard to create on aircraft. Flush rivets and welded seams need be used only if convenient, not for any streamlining effects.

There is no need, either, for even "pointing" a long, thin object at its projected destination. Intraspace vehicles might well travel "sideways," in spite of *2001* predictions to the contrary. In fact, there's no reason to suppose that an intraspace vehicle would be long and thin at all. A square block, a round globe, or an equilateral-sided pyramid might be better suited to the needs of intraspace transportation. Most of the projected design concepts for space stations envision them as round rings. Any design you see today is someone's "best guess" and, while some of those guesses may be very well-educated ones, even our computer technology is only as good and imaginative as its human programmers.

One of the endearing aspects of ETV modeling is that it offers boundless opportunities for true artistic creativity. That square block vehicle might be probable, but man, given the freedom of design space provides, will probably come up with something that also has visual appeal. The unwritten law that implies "what looks

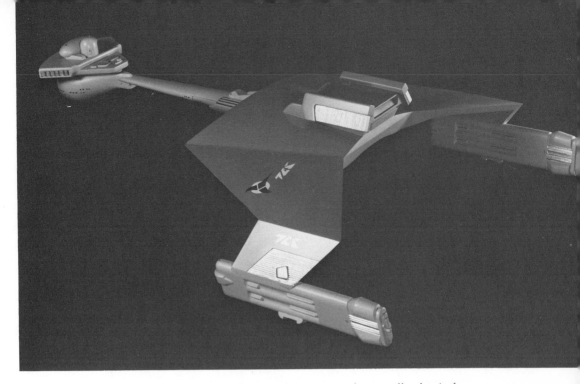

Fig. 7-2 The Klingon Battle Cruiser is the equally classic foe for the Star Trek™ Enterprise. AMT has two different size kits for both ETV models. *Courtesy AMT Corp.*

Fig. 7-3 Flying model kits are offered by Estes for both stars of the "Star Trek" television series. *Courtesy Estes Industries.*

right is right" will probably save us from seeing truly ugly vehicles in actual service. It's far more likely that the actual intraspace vehicles will be at least as appealing as those in *2001: A Space Odyssey* or *Silent Running*.

There's always the unpredictable advance or breakthrough in technology that could alter the shape of intraspace vehicles. Solar power, for instance, could alter our ideas of bulging pencil shapes to pancake-like conceptions for such vehicles. The two types of T.I.E.-Fighters in *Star Wars* and the Ziggurat proposal in this chapter are examples of possible intraspace vehicle shapes where solar power is a consideration. The only real constriction in the design of an intraspace vehicle, then, is that it at least appear to be functional.

Designs for several original spaceships are given in this chapter. The Battle Corvette Nebula is a fairly complex-appearing model that is incredibly simple to build. It would make as good a first time scratchbuilt model as you could find, and, thanks to the type of de-

Fig. 7-4 The Battle Corvette Nebula is a 1/120 scale model you can assemble in a few evenings with the full-size patterns.

tail parts used, you'd have a choice of building one anywhere from a foot long to about four feet long. Experienced modelers can start right off with the Battle Carrier Roo or the Battle Cruiser Passim. If even the Nebula seems a bit complex, then try just one of the Petal Class Fighters before going on to the Nebula itself. You might even want to create your own fleet of adversary ships based on the solar-powered Ziggurat.

The Battle Corvette Nebula

The Battle Corvette Nebula was created especially for this book to provide an example of the form-follows-function design philosophy and to provide you with at least one alternative to the seemingly endless proliferation of television and motion picture rocket models. The basic structure for the Nebula is the triangle, a sturdy, efficient, and economical structural shape.

Nebula incorporates another design feature that is likely to appear on many future ETV miniatures, namely it has separate systems for intraspace and interspace travel. The nose structure is designed for reasonably efficient travel in zero atmosphere and zero gravity but with built in adaptability to stable flights, including the ability to glide or soar, in an atmosphere and gravity environment similar to Earth's. This nose section is detachable from the Nebula, and both the rear or intraspace portion of the ship and the nose section can travel completely independently of one another. The nose section could, in theory, provide enough flight stability to allow a full-scale Nebula to land on earth, particularly with the wings or "petals" of the nose section extended in the delta wing position.

Nebula is envisioned as a fantasy war game vehicle so that the detachable nose section would, logically, be a fighter, one of the Petal-Class Fighters, in fact. The Battle Corvette Nebula, then, has its own elaborate defense systems including laser cannons and other smaller firepower as well as the immediate capabilities of its own Petal-Class Fighter protection.

Designing is the first step in the creation of any custom space vehicle so you can make corrections simply with an eraser. I se-

Fig. 7-5 Designer's conception of two of the Nebula-Class Battle Corvettes, one with its Petal-Class Fighter just launched and the other with the fighter still docked in the ship's nose.

Fig. 7-6 Four Petal-Class Fighters escort the Battle Cruiser Passim and its three docked fighters on an intraspace mission to search out invading Ziggurat Ships.

Fig. 7-7 Five 1/72 scale kits are needed to make the Battle Corvette Nebula.

lected an overall size for the model based on the relative bulk of 1/72 scale tank kits and their detail parts. The actual model is approximately 1/120 scale. There are dozens of 1/72, HO, OO, and 1/76 scale tanks, guns, and other armored vehicles available in the larger hobby stores. Most of these are also available in 1/48, 1/35, and 1/32 scales. If the foot long model to match the templates isn't large enough for you, then just double the size and use parts from 1/48 scale tanks, or make the model four times the size of the patterns (about four feet overall!) and use 1/32 or 1/35 tank parts.

I used engines from the MPC brand Space: 1999 Hawk. AMT brand Star Trek Enterprise engines could be used for the larger version, or you could select a suitable Estes or Centuri set of flying rockets for the engines. Do not, however, try to make this one fly—it's only intended to be stable in zero gravity!

Make the three double-triangle body pieces and the single triangle base section from .020 inch thick Evergreen styrene plastic sheet stock (or from one of those plastic "For Sale" signs). Note that the same pattern is used for the three body sides as for the base, *except* that the base has three corners cut along the dashed lines. The inner, intermediate, and outer triangles or "petals" are strictly decorative unless you want to make a "wings-out" petal fighter like that shown in Chapter 6. These parts, too, should be cut from the .020 inch thick plastic. Use the slice-and-break method to cut the parts, and then file the edges smooth.

Attach some 8606 6 x 6 HO-scale Evergreen brand styrene strips along one edge of each of the double triangle body sides and along the base edge of each of the body sides (see figure 7-14). File the edge of these reinforcing strips at a 30 degree angle using a medium cut flat mill file (figure 7-15). The first two sides can now be glued together, using the base piece to be sure the angle is 30 degrees (figure 7-16). Add the third side and finally the base. Tube cement is best for all this assembly because of the length of the glue joints.

Two 1/72 scale Crusader Tanks and two 1/72 scale Tiger Tank kits were used for the surface details with some additional half-round tank supports, clear plastic bubbles, 3/64 wire-reinforced round Plastruct brand parts, and the Space: 1999 Hawk kit for the three rocket pods. The tops of the hulls of the two Centurion Tanks were cut in half using the PanaVise and a block of oak as a support. The gun barrels on the Tiger Tanks were cut shorter. All the other parts were simply glued in a more or less symmetrical pattern to the sides of the model. One of the three sides of the body was designated as "bottom" and it was detailed to represent various exhaust jets that might be used for stabilizing and in-flight maneuvers.

The Petal-Class Fighters

The Battle Fleet that includes the Corvette Nebula was designed with the Petal-Class Fighter as an integral part of each ship. The single Petal-Class Fighter forms the nose of the Battle Corvette Nebula, 19 Petal-Class Fighters form the nose and the surface of the Battle Carrier Roo, and three of them form the nose of the Battle *(Text continued on page 126)*

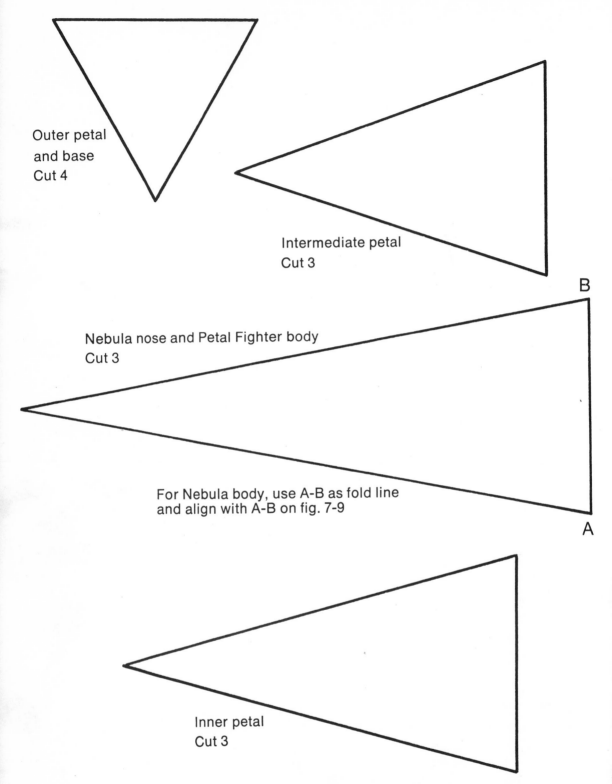

Outer petal
and base
Cut 4

Intermediate petal
Cut 3

Nebula nose and Petal Fighter body
Cut 3

B

For Nebula body, use A-B as fold line
and align with A-B on fig. 7-9

A

Inner petal
Cut 3

Fig. 7-8 Patterns for a 1/120 scale version of the Battle Corvette Nebula nose section and the Petal-Class Fighters.

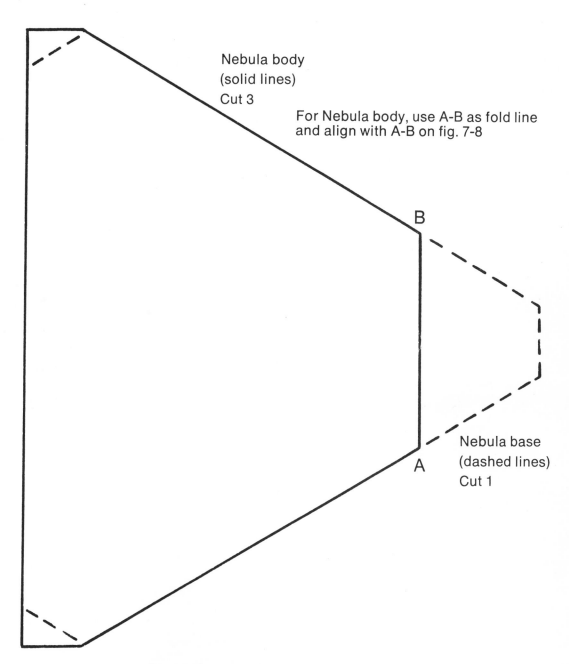

Nebula body
(solid lines)
Cut 3

For Nebula body, use A-B as fold line
and align with A-B on fig. 7-8

B

Nebula base
(dashed lines)
Cut 1

A

Fig. 7-9 The body section of the Battle Corvette Nebula
needs three pieces the size of the solid lines and one the
size of the dotted line pattern for a 1/120 scale model.

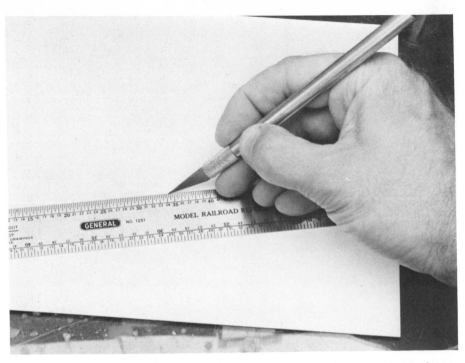

Fig. 7-10 Copy the pattern and trace it onto a sheet of .020-inch-thick plastic. Slice lightly along the lines with an X-acto knife guided with a steel ruler. There's no need to cut clear through the plastic.

Fig. 7-11 Snap the plastic over the edge of a table or a piece of Plexiglass to break the plastic cleanly along each of the light cuts.

Fig. 7-12 The three sides of the Battle Corvette Nebula can be single pieces with a bend at the base of the slender front triangles.

Fig. 7-13 File each of the edges of the three side panels with a large size medium cut mill file.

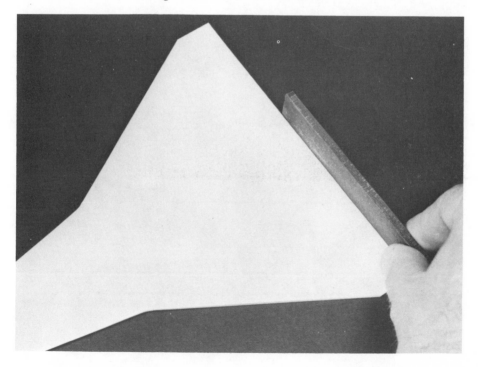

Fig. 7-14 Cut some Evergreen brand 6 x 6 HO scale styrene to serve as interior braces along all the seams of the Battle Corvette Nebula. Glue the braces to the sides with liquid cement.

Fig. 7-15 File the glued on braces at a 30 degree angle to fit snugly inside the adjoining side when the three side pieces are assembled.

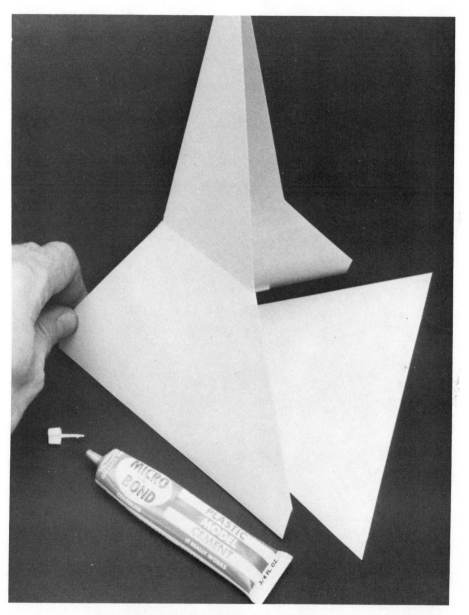

Fig. 7-16 Use the single base piece as a guide for gluing the first two sides together so you know the angle between them is correct.

Fig. 7-17 Glue the second side in place with tube type cement and then glue on the base piece. If some of the cement dries before the parts join, just brush on some liquid cement for plastics.

Fig. 7-18 The unpainted Battle Corvette Nebula shows exactly where the various tank parts are located. Paint will effectively disguise the origin of the parts.

Fig. 7-19 The third side of the Battle Corvette Nebula has a different set of tank parts. Most of the exhaust nozzles on the base of the model are wheels from the four tank kits.

Fig. 7-20 The model is painted, decals are applied, and then it is weathered with a coat of Testors Dullcote. Finally the Plastruct clear plastic domes are glued in place with rubber cement.

Cruiser Passim. All of these Fighters are designed to be ejected for defense in case of attack or offensive maneuvers.

The Petal-Class Fighters utilize folding wings or "petals" to provide the streamlined coverings necessary for atmospheric travel. A view of the environment surrounding the vehicle is provided via various sonar, radar, and laser devices, with auxiliary slit windows, similar to those on tanks, in the edges of the innermost petal. The two outer petals can be pivoted outward, at the rear of the vehicle, to allow the rocket engines they cover to be used for reverse thrust. The Petal-Class Fighters return to their mother ships with the help of this rearward mobility. These fighters rely on their maneuverabi-

Fig. 7-21 This model of the Petal-Class Fighter has only the outer and intermediate petals. Both are open for a reverse thrust maneuver.

lity, including almost instantaneous reverse through the use of the petals and reverse rockets' thrust, for defense. A single metal-seeking laser cannon fires from the tip of the fighter's nose. The simple exterior structure also provides for effective armor style shielding and force field orientation around the tips of the triangular surfaces.

The outer two petals on two sides of the three-sided fighter are also hinged along their longest edges to serve as wings for atmospheric flight. These two panels are folded outward and locked in position. Interior flaps are used to vary the amount of lift for landings on planets with atmospheres. The fighters are equipped with conventional folding landing gear with sponge rubber tires for constant shock absorbing qualities regardless of the atmospheric pressure. The Petal-Class Fighter was built as an integral part of the 1/120 scale model of the Battle Corvette Nebula because it was easier to

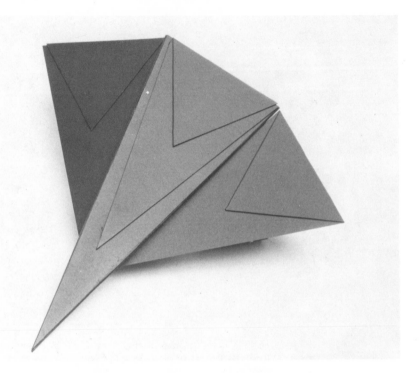

Fig. 7-22 The Petal Class Fighters can unfold their outer and intermediate petals for soaring flight and battle maneuvers in atmospheric conditions.

make one-piece sides with a fold rather than a seam. The patterns will allow you to build the model either way and to make additional Petal-Class Fighters if you wish. You may want to assemble one or two of the fighters with their wings or "petals" in the position for atmospheric flight (the delta-wing configuration) or folded out for reverse thrust to dock on the pockets of the Battle Carrier Roo. The Petal-Class Fighter portion of Nebula was painted with an aerosol can of Pactra's Navy Blue, the body of Nebula is Pactra's Rebel Gray. The weathering is a combination of a quick pass with a spray can of Floquil's Engine Black with additional streaks of Engine Black sprayed through a cutout in a paper mask (see Chapter 3) and a final wash of water and black acrylic. The decals are from the Centurion kits and were applied before the weathering. A final light coat of Testor's Dullcote was applied just before the clear plastic bubbles were glued in place with five minute epoxy. Be careful with clear spray paints; almost all of them will cloud or etch clear plastic!

The Battle Cruiser Passim and Battle Carrier Roo

The Battle Cruiser Passim is an example of its class and the largest ship in the fleet to carry its own defense systems. The Passim (Latin for "at different places") carries a total of three Petal-Class Fighters as detachable nose sections. It is shown in figure 7-6 with an additional defense complement of four fighters from one of the Battle Carriers.

The Battle Carrier Roo is the mother ship for a total of 19 Petal-Class Fighters. The outer surfaces of the fighters form the outer skin of the carrier when they are docked in their pens or pods. A single fighter doubles as the nose section of the Roo, three more are carried just below the nose, six more below that, and nine more fighters are carried near the main section at the rear of the ship. All 19 fighters are positioned for instantaneous launching, but the fighters must unfold their petals to reverse direction for docking on the mother ship. The surface of the Battle Carrier Roo is sealed and armored so the ship can maneuver when all of the fighters are deployed.

The Ziggurat

The Ziggurat is a proposed adversary for the Nebula and its fleet. Ziggurat is designed around an immense square panel of solar energy cells. That panel would be faced toward the nearest sun during cruising maneuvers, but there is storage capacity to allow battle maneuvers in any position. The fighters in Ziggurat's fleet would share a triangular shape similar to the Petal-Class ships but with sides of unequal length. Ziggurat's fighters would also form the nose section of the ship.

Figure 7-26 shows a pattern for the unusual square-on-triangle shape of the Ziggurat with fold lines and letters that are to be

(*Text continued on page 138*)

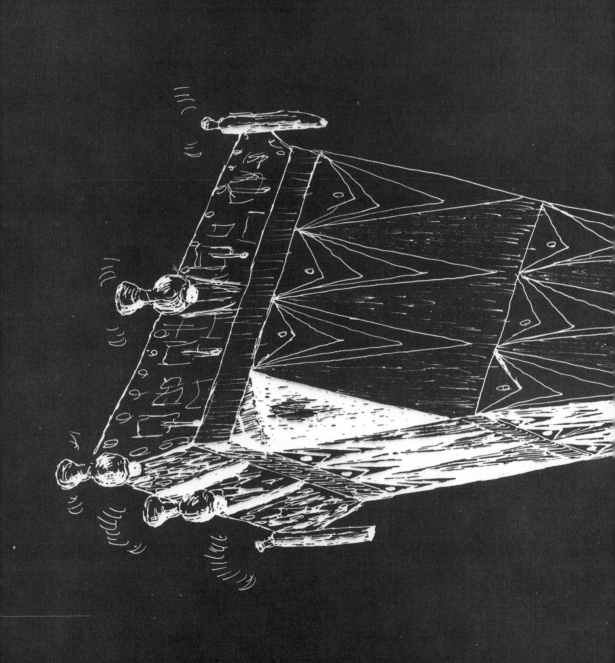

Fig. 7-23 Designer's sketch of the Battle Carrier Roo with all but one of its 19 Petal Class Fighters docked. The 19th fighter has its petals open and is docking.

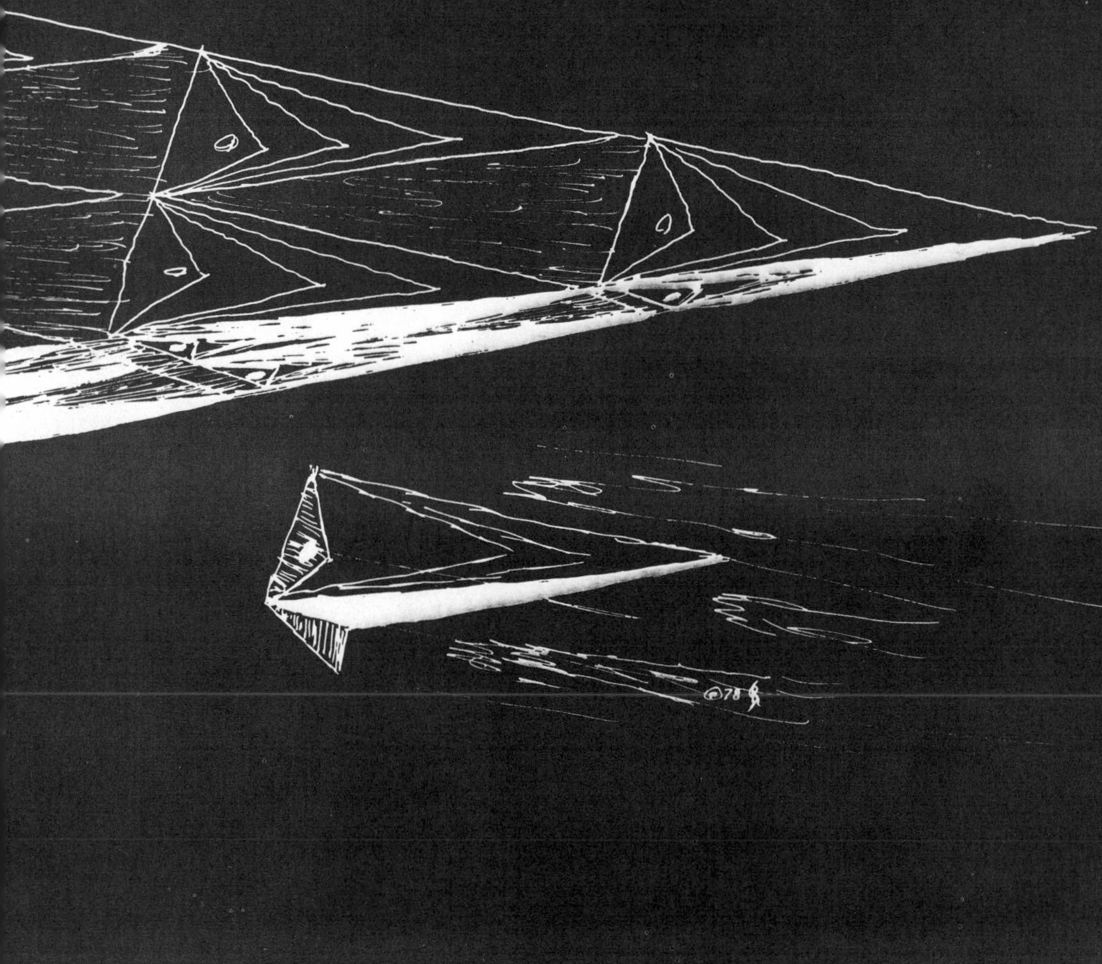

Fig. 7-24 The Roo-class battle carriers can defend themselves and maneuver with all 19 Petal Class Fighters out of their bays.

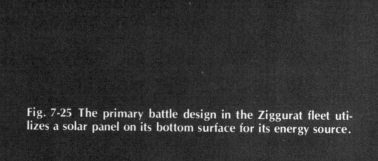

Fig. 7-25 The primary battle design in the Ziggurat fleet utilizes a solar panel on its bottom surface for its energy source.

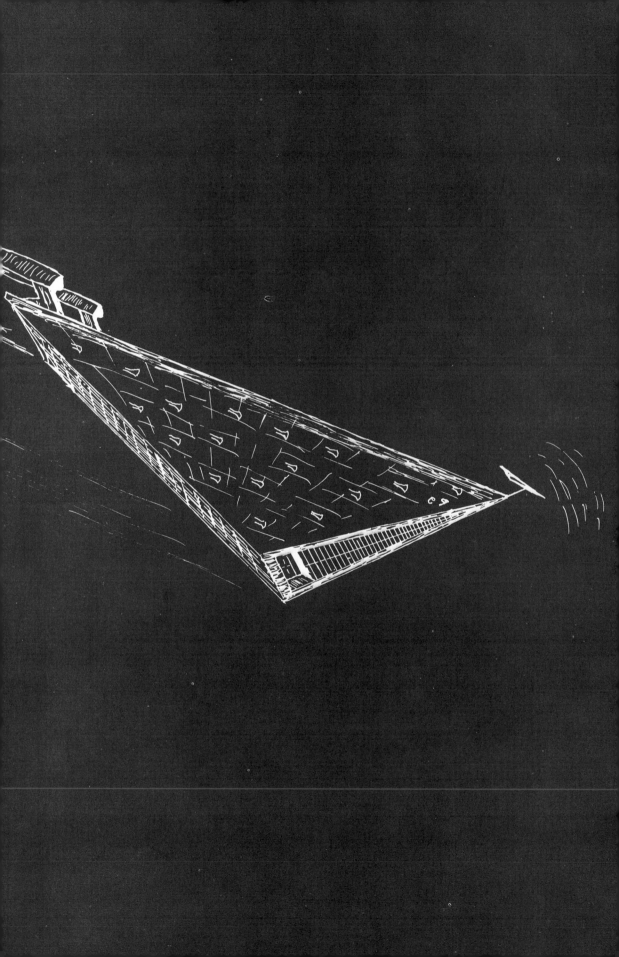

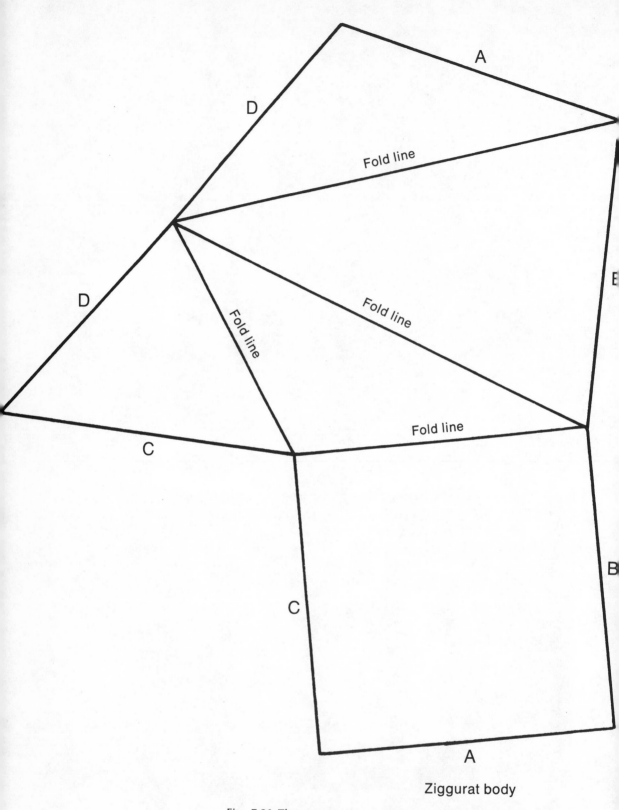

Ziggurat body

Fig. 7-26 The pattern for the Ziggurat intraspace vehicles. Edges A, B, C, and D must be joined to their matching letters after the part is folded inward on each of the fold lines.

Fig. 7-27 A side view of the design study of Ziggurat. This one was made from .020-inch-thick sheet plastic.

Fig. 7-28 A top view of the Ziggurat design study reveals the relative position of the square bottom solar panel.

matched together on joining surfaces. The dimensions of the various sides can be enlarged as much as needed for a model of Ziggurat as long as every line is enlarged the same amount. A thin slit should be cut all the way across the rear edge of Ziggurat to simulate the main rocket engine's primary exhaust ports. The two rocket engines on pods are adjustable for additional maneuvering force. They are of appropriate size to be parts of a model rocket kit.

Ground Effect Vehicles

The "future" has been with us longer than you might imagine. Vehicles that fly with "force fields" have carried thousands of passengers for more than a decade. The force fields are compressed air ducted toward the earth by a skirt around the outer edges of the vehicle. These machines are called hovercraft or ground-effect vehicles and they quite literally float on air. The hovercraft uses its huge horizontal fan to move enough air to keep it away from the ground, or, in the case of the practical applications of hovercraft, away from the water. The principle is even used by some cannister-type home vacuum cleaners where the exhausted air from the vacuum is ducted beneath the cannister so it will "float" over the carpet.

The ground-effect vehicle has but one unusual limitation for a "flying" craft: its flexible (or telescoping) "skirt" must be in touch with the ground to provide the sealing effect necessary for enough air force to keep the vehicle afloat on its blanket of air. The hovercraft vehicles hang a flexible skirt around the edges of the fan that generates their air pressure. It's the partial seal provided by the skirt that makes the pressure on the air generated by the fan effective. The hovercraft literally floats on a layer of air pressure greater than that of the surrounding atmosphere. Most full-size hovercraft resemble the Queen of Hearts from *Alice In Wonderland* more than they do a sleek spaceship, thanks to that bulging, flexible skirt around their edges.

The hovercraft's movement is definitely one of a true floating vehicle with none of the sudden rocking and leaning typical of a tracked vehicle like a tank or a wheeled vehicle such as a dune

Fig. 8-1 A profile view of the Testors Galax IV ready-to-fly hovercraft.

buggy. This is because the air buoys the vehicle up and also acts as a superb lubricant between the hovercraft and the surface beneath it. The result is an incredibly smooth movement marred only by a slight bobbing effect as the air pressure escapes when the skirt lets a bit too much out at a bump or the peak of a wave. That floating effect is eerie to observe because, until anti-gravity does arrive, there's no other kind of vehicle on earth that moves quite the same way.

Hovercraft Models

The hovercraft or ground effect principle is far better suited to a model than to a full size vehicle because the model can be much lighter, even in relative terms, than the real thing. The "bubbles" of air that form beneath the fan of a hovercraft are much closer to the size of a model, making it easier for the model to float. In fact, the ground-dragging skirt of real hovercrafts isn't really necessary if the model is constructed properly. To define "properly" for the con-

Fig. 8-2 The Galax IV "flies" about a quarter inch above the ground on a cushion of air generated by its own glow fuel-burning engine and propeller.

struction of a model hovercraft, take a close look at some of the ready to hover "toys" now on the market.

Galax IV's Force

The Testors Corporation has a 16-inch-long ETV-style model called the Galax IV. This vehicle is powered by a model airplane's glow fuel-burning two-stroke engine with .049 cubic inches of displacement. A large propeller is mounted on the engine, and the engine is mounted so that the propeller revolves horizontally to push air out the bottom of the Galax IV. The relatively high speed of the propeller and the fact that it is closely surrounded by a plastic skirt allow it to produce a very effective layer of high air pressure beneath the vehicle.

The Galax IV is self-supporting on its own air pressure as long as the surface beneath is it smooth. The vehicle hovers nicely over relatively flat concrete, blacktop, or even dirt. Dirt can be blown back into the engine, however, so it's best to limit operation to paved areas.

There are no directional controls provided for the Testors model, so it propels itself away from any "leaks" of air in a completely random series of movements. It's a toy only because it's uncontrollable. Its flights, however, are virtually a match for the eerie bobbing and ultra-smooth movements characteristic of hovercraft.

The Galax IV could be an ETV fan's experimental paradise because, as a toy, it's relatively inexpensive. Two or three of the Galax IV engine, propeller, and duct units just might be enough to support a lightweight radio control receiving unit and battery. The R/C components could be borrowed from one of the small and inexpensive radio control automobiles. The radio control units would allow you to control the path of your Galax IV-based vehicle from a distance.

The direction of the Galax IV is determined by the slightest amount of lean of the vehicle. The radio control gear could be rigged to move a small weight on the end of a long arm inside a "body" built around the propeller/engine units so it would not be visible. If the weight were moved toward the front of the vehicle, the vehicle would tilt forward slightly and allow a bit of the ducted air to escape out the rear. Forward travel would result. The weight would have to be adjustable on the arm to allow for trimming to keep the vehicle from tilting too far. Model airplane shops sell small clamp-on collars that could be used with a simple wrap of wire type solder around a plastic or steel rod on the radio control servo mechanism.

An alternate method might be to use a small (about 2-inch square) piece of aluminum that could be moved over the top of one of the two or three fan ducts, by the R/C servo motor, to reduce that engine's power. The system would probably be most effective if the servo and the damper flap were mounted in the center of three Galax IV propeller/engine units so the damper could be moved clockwise with the servo motor to partially cover any one of the three units or none at all.

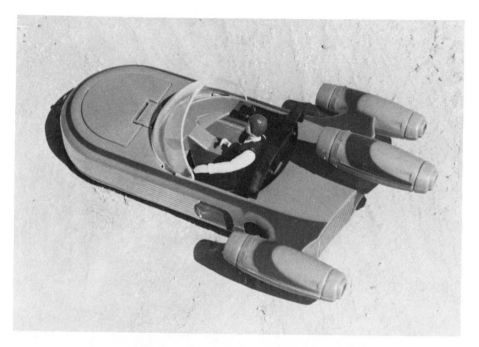

Fig. 8-3 The Kenner Star Wars™ toy version of the Land Speeder™.

The power unit from the Testors Galax IV should open up a whole new category of ETV action models to any imaginative modeler. There's no reason why the vehicle must be under your complete control. The random flights of the Galax IV right out of the box are a delight and, if you really do want it to go off in some particular direction, all it takes is a light prod with a foot or stick. It's as close as you're likely to get to the *visible* performance that an anti-gravity machine would have!

You can, then, use the power system from the Galax IV to create your own scale models of zero-gravity vehicles. The only difficult part of any such project will be in keeping the body you build as light as possible. The best material is probably balsa wood in sheets or hollowed out beams, with as few coats of sanding sealer and paint as possible. The methods and materials used to construct flying model gliders with balsa ribs and a paper or Silkspan covering

Fig. 8-4 The Ballantine Books packets of blueprints contains plans for both flying and surface vehicles. The paper punch out models from the Random House book, shown at the top of the picture, could be used for patterns for larger scale working models.

would also work well. In fact, you might well find a model airplane kit that could be adapted to a hovercraft design by, for instance, splitting the fuselage down the middle. With some care and attention to weight, you should be able to build a body as large as the Galax IV but to any design that suits you.

Ballantine Books has a set of *Star Wars Blueprints* that includes plans for the Land Speeder. If you're experienced with model airplane construction techniques, you should be able to build a 12-inch-long model of the Land Speeder that looks like it has the same anti-gravity force as the vehicle in the motion picture.

You will have to experiment with a position for the Galax IV engine/propeller unit to keep the Land Speeder balanced properly in spite of the bulk of its three rear engines. The model would have to be finished with the exception of balsa mounting braces (or a $1/16$-inch balsa sheet floor) before the correct position for the propeller/engine unit and its round duct were located. Even then, it would be wise to mount the unit with screws and perhaps with adjustable aluminum strip brackets so additional adjustments could be made after some test flights. Most of the area behind the Land Speeder cockpit would have to be cut away or covered with open louvers to provide the air intake for the propeller. It would be best, too, to use a smoked clear plastic bubble over the cockpit like that in the plans rather than the convertible cockpit of the movie vehicle. The bubble cockpit would allow you to skip any interior details including, of course, any of the Kenner Star Wars dolls which would be far too overweight for the model.

CHAPTER 9

Space Buggies

The concept of extraterrestrial vehicles must include the more mundane machines that will be needed to provide surface transportation and other services on the surfaces of moons and other planets with less gravity than our own. One could safely assume that Mack truck-like machines would be used to cope with planets or moons having gravity in excess of our own. Since the only travel so far considered has been on the reduced gravity surface of the moon, very little has appeared in either the fictional or factual press regarding vehicles with multiple gravity capabilities. The various aerospace and other industrial firms that bid on the Lunar Roving Vehicle contract presented a variety of solutions to the problems of mobility on a powdery, soft surface with less than one G. Those vehicles and the vehicles created by the *Star Wars* crew are only the beginning of man's imaginative proposals for surface travel in alien environments. It's another wide open area for creative modelers to explore with combinations of kits and scratchbuilt components.

Radio Control Magic

The concept of remote control lends itself especially well to extraterrestrial surface vehicles or space buggies. It's highly likely, for one thing, that such vehicles will, indeed, be remote control so that space explorers can do much of the work from the protection of a base station. If you've ever operated a remote control car or tank, you've experienced the haunting sensation of seeing a machine that is ordinarily driven by an on-board human perform by itself.

146

Radio control has been around for decades as a remote control system for model airplanes. "R/C," as it's termed, has also been available for some time at relatively low cost in model racing automobiles. It's only been affordable for most of us, though, for the past three or four years. Radio control is one of the newest hobby areas because it is just now receiving the benefits of miniaturization from the aerospace and computer industries.

A radio control system is essentially just an electric motor, called a servo, on the vehicle that is powered by batteries that also must be carried along. A small radio wave receiver unit turns the motor on or off in response to commands from the remote transmitter. The transmitter is the control box you hold in your hands. Most of the systems used for flying model airplanes require the simple formality (your dealer can do most of it for you) of getting a Federal Communications Commission (FCC) license to transmit. Many of the inexpensive systems don't even require that.

Automobiles and even robots like Kenner's Star Wars toy radio

Fig. 9-1 The Kenner Star Wars™ toy version of R2-D2™ moves forward or backward and turns left or right in response to push-button commands from the transmitter.

Fig. 9-2 The Cox dune buggy is powered with a glow fuel-burning model airplane style engine. It's available with radio control steering or as a free-running model.

control R2-D2 are available for about $30 or so with two channel control. The two channel system provides two controllable functions such as left or right steering on a model automobile. In the case of Kenner's miniature, a "head" rotates to determine the direction in which you want the robot to go; you also have the ability to start and stop his motion. The Kenner Star Wars toy R2-D2 will go forward or backward and turn left or right.

The dune buggy you see in figures 9-2 to 9-4 is a Cox ready-built model powered by an .049-cubic-inch-displacement, two-stroke, glow fuel-burning model airplane engine. Cox offers the vehicle with a radio control setup, or you can do as we did and adapt one from one of the radio control cars. We used the unit from the Jerobee brand racing cars because we wanted the second channel speed control. Frankly, with a dune buggy, the speed control isn't

Fig. 9-3 This Cox dune buggy has been fitted with the fully proportional steering and carburetor (acceleration) servo from Jerobee.

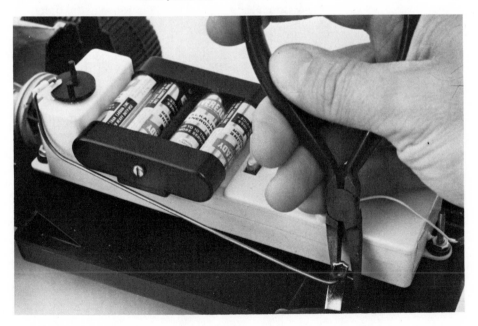

Fig. 9-4 Goodyear Pliobond will help to make the bolts and nuts vibration proof on a radio control model.

used all that much, and the Cox ready-to-run R/C dune buggy is good enough.

Both the Cox system and our converted vehicle have "fully-proportional steering," a fancy R/C term for steering or other function control that works faster or turns sharper the harder you push on the control stick. A special carburetor was needed, in addition to the R/C receiver, to allow radio control of the dune buggie's acceleration as well as its steering.

There are considerable differences among different radio control systems. The words to look for when buying any R/C item are "fully-proportional" and "channels." Some of the servos in the receivers are just on–off solenoid-type devices that position a lever for steering, turns, or whatever. The more expensive servos have motors that move the lever over a fairly wide range so that a dune buggy, for instance, can be controlled to turn tight or wide radius turns and anything in between. This infinite degree of control, from idle to full-throttle on a carburetor controlled by a servo motor, is called "fully proportional."

Obviously, it costs more to manufacture a motor than a simple relay, so the fully proportional systems are more expensive. The fully proportional systems are also heavier because of the motor and the larger batteries. That extra weight can be a detriment in a vehicle like the hovercraft described in the previous chapter.

As a rule, the inexpensive ($20 to $60) radio control vehicles have nonproportional control and often just one or two channels. The channels are the number of servos or solenoids on the receiver and the number of matching controls on the transmitter. You can buy a car or a tank, for instance, that will steer left or right on only one channel. Some designers are even clever enough to get a forward and reverse or a simple slow–fast speed control from one channel by using "over-center" controls or special linkages inside the vehicle (like those in Kenner's Star Wars toy R2-D2 radio-control model). For most space vehicle applications, one or two channels are quite adequate, and there's little need for the expense or weight of the fully proportional feature.

Wheeled Vehicles

The Lunar Roving Vehicle (LVR) that went to the moon with Apollo 17 was a wheeled vehicle, as were most of the other design proposals for the first vehicle to traverse another planet or moon. The LVR looked like a complicated four wheeled motorcycle because it lacked a body. Some of the other proposals did include bodies, and some were even rather stylish. It's doubtful that there will ever be a "lunar rover" like the MPC brand kit for the 1/24 scale Space: 1999 Alien. It's a bit surprising, though, that there were so few LVR design proposals with the integral monocoque body and chassis that have proven so successful on Grand Prix and Indianapolis racing cars. If you want to design a better LVR, then the suspension and tires from the Alien would be one item to start with. If you want to come a bit closer to the real LVR suspension, then start with one of the 1/25 scale Indianapolis or Grand Prix plastic car kits with, perhaps, the wheels and tires from a similar-size plastic jeep or other military vehicle kit.

Tracked Vehicles

So far there have been only a few suggestions for ETV surface machines that are propelled by tracks like a tank's rather than by wheels and tires. The gigantic Sandcrawler™ in *Star Wars* was a tracked vehicle that in reality started out as a model tank. You might be surprised to learn that Star Wars R2-D2 was really a tracked vehicle. The little models of the robot figure usually replace the tracks with wheels because, on a small scale, wheels are more practical. But if you were to build a 1/4-scale (about a foot high) replica of R2-D2, there would be enough space in each of the three "feet" for the complete mechanism of one of the inexpensive radio-controlled

Fig. 9-5 The wheels, tires, and suspension from MPC's 1/24 scale Space: 1999 Alien space buggy kit (© 1977 ATV Licensing Ltd.) would serve nicely for freelance space buggy miniatures. (MPC kits are products of Fundimensions, CPG Products Corp.) *Photo by Charles Hepperle.*

tanks. You'd need to find some that could be operated on three different frequencies to provide independent radio control for each "foot." The principal dimensions for the model can come from the MPC-brand kit for the robot.

A working radio control model of the Star Wars Sandcrawler could be constructed from four such R/C tanks, or you could build just a display model from static model plastic tank or half track kits. Plans for the track portion of the Sandcrawler are included in the Ballantine Books' packet of *Star Wars Blueprints*. A single side view of R2-D2 is also part of the packet, as are four other surface vehicles that did not star in the first film. The principal Sandcrawler dimensions can be determined from photos, using the plans for the treads for the proper proportions.

CHAPTER 10

The Space Scene

The incredible realism achieved in motion pictures is partially the result of placing the extraterrestrial vehicles in a total environment. It is the backgrounds as much as the models themselves that make those movie and TV space scenes seem so real. The photography and special effects for exhausts and weapons fire add the final touch. We'll talk a bit about some simple photographic techniques in the next chapter. For now, let's see what can be done to give an ETV miniature some of the background it deserves. Your friends will find your models far more realistic if they don't have to imagine the background and other special effects that should surround the model. You'll even improve your own image of your model building by converting the background space scene from imagination to three-dimensional reality.

The Creation of a Planet

You certainly do not need to build an entire planet or a moon or even a small satellite to provide a realistic background for your ETV models. All you really need is enough of the surface of that planet, moon, or satellite to completely frame the model.

A scene about 2 inches or so larger than the width and length of the vehicles you want to display is ample. What you really want is to duplicate the effect of the frame around an artists' oil painting. You want to create a "window frame" that gives the viewer the feeling that the scene continues on to infinity beyond the boundaries of the diorama. Both your model and your background can be three-

154

Fig. 10-1 This is Robert Angelo's 1/48 scale diorama assembled from Minicraft/Hasegawa model aircraft figures and accessories and MPC brand kits of the Star Wars™ X-Wing Fighters™. (MPC kits are products of Fundimensions, CPG Products Corp.)

dimensional, but you can also choose to combine three and two-dimensional art forms into a single scene.

One of the most effective dioramas is the shadow box, which is essentially a picture frame that surrounds a three-dimensional scene to give the appearance of a painting. A number of natural history

Fig. 10-2 Robert Angelo used leftover scraps from aircraft model kits to build this ground support vehicle for his hangar diorama.

museums use room-size shadow boxes to illustrate the environments of prehistoric animals or historic events. These dioramas may
be 1/1 scale (full-size) or they may be miniatures on the scale of
most model kits. You can have a shadow box without a diorama or
a diorama without a shadow box, but the combination of both is
better than either.

Robert Angelo's hangar scene with Star Wars X-Wing Fighters
(figure 10-1), Kenneth King's engine display (figure 10-3), the MPC
brand model of the Space: 1999 Alpha Moonbase (figure 10-4), and
Estes' moon-like rocket display (figure 10-5) are all dioramas. If any
of them were mounted behind a picture frame with some kind of
painting or photo mural (or just black velvet to simulate space) they
would become shadow box dioramas. The black lit scene with the
two Star Trek vehicles (figure 10-7) is an example of a shadow box
and, because of the stars and planets on the backdrop, it also quali-
fies as a diorama. This one would, obviously, look much better with
a picture frame disguising the edges of those corrugated cardboard
cartons.

Fig. 10-3 Some Minicraft/Hasegawa models include detailed
engines, like this one from their F-86 fighter, that Kenneth
King uses to add extra realism to rocket hangar dioramas.

Fig. 10-4 The injection-molded buildings and the vacuum-formed plastic base in the MPC brand Space: 1999 Alpha Moonbase (© 1977 ATV Licensing Ltd.) will work nicely with any medium-size ETV model to make a diorama. (MPC kits are products of Fundimensions, CPG Products Corp.)

The in-flight scenes like the Star Trek shadow box are the type of display many will consider first, but I honestly cannot recommend them. The Star Trek scene photographs nicely (see figure 11-1) and it looks fantastic under black light, but those are not conditions you can actually live with. For day-to-day enjoyment, a static display with the model docked on an artificial asteroid, on a simulated planet, or in a hangar are much more realistic. But do give the black light diorama a try. It's relatively inexpensive because you'll probably only have to buy the paints and one or two black light bulbs, and the visual experience of seeing those models floating in space (if not actually in flight) is breathtaking.

The military and aircraft model kits should have all the figures you need for just about any hangar or dock scene. The MPC brand

Fig. 10-5 Special texturing material can be added to common patching plaster to produce a rough surface like this lunar diorama. *Courtesy Estes Industries.*

Star Wars vehicles are 1/48 scale, so you can use mechanics and ladders from the Minicraft/Hasegawa aircraft models as well as their various fighting figures from the military series. There are other brands of 1/48 scale figures and similar ones in 1/76, 1/72, and 1/35 scales.

The various 1/160 (N scale), 1/120 (TT scale), 1/87 (HO scale), 1/76 (00 scale), 1/64 (S scale), and 1/48 (0 scale) scale model railroad buildings, vehicles, and people are also suitable for ETV dioramas. The model railroad shops will also have all types of detail castings for matching the cluttered effect of intraspace vehicles' surfaces. Some of the kits for plastic oil tanks, oil filling facilities, and bridges will also have parts that can be used, in any scale, for ETV dioramas. The model railroad shops will also be able to supply virtually every imaginable type of construction texture, from bricks and shingles to stones, in plastic sheets. For concrete, you can use either plain plastic "For Sale" signs or, better, a thin coating of plaster over 1/2-inch or thicker plywood.

While you're at the model railroad shop, check out the Mountains-in-Minutes brand of expanded polyfoam plastic scenery kits. Expanded polyfoam is mixed as a fluid and expands as it cures to a texture a bit like a hard sponge but is much easier to cut. The expanded foam's surface looks very volcanic or other-worldly if left as is, or it can be used with that firm's rubber molds to make earthly rock textures.

The material easiest to find for making replicas of the rugged lunar surface or for sculpting the surface of a planet is plain old patching plaster sold by hardware stores. You can mix it to a soupy consistency and dip paper towels in it for self-supporting scenery. Just build your extraterrestrial landscape with wadded-up newspapers and drape the plaster-soaked paper towels over it. You can brush or trowel-on a cover coat of plaster to get the rough effect you desire.

Fig. 10-6 Two large cartons must be taped and glued together as shown for the black light Star Trek™ diorama. The strips at the bottom will hide the light itself.

Fig. 10-7 The completed black light shadow box/diorama with tape to hide the rough cardboard edges. The light is in place behind that lower panel.

Paint and hardware stores sell a variety of materials from spackel to ground-up walnut shells of various sizes to give walls and floors a rough texture. Just mix the rough texture material in with the patching plaster for that final coat. The more of the texture material you use, the rougher the surface will be. You can mix almost equal parts plaster and texture material, with some brands, to achieve a surface like that in the Estes diorama (figure 10-5).

Black Lighting Effects

Black light (ultraviolet light) was very popular as a room decorating effect a few years ago. Hundreds of posters appeared that were printed with ink that would respond to ultraviolet light. Many of the shops that sell the posters also sell ultraviolet light bulbs and bottles of special paint that is sensitive to ultraviolet light. Many hobby craft shops also sell the paints. The ultraviolet light bulbs are available in an incandescent (screw-in) bulb and in fluorescent tubes. Two-foot fluorescent tubes mounted in a $20 desk-top fixture are best for a shadow box the size of the one in figures 10-6 and 10-7. A single fluorescent tube or one of the light bulbs would be large enough for a shadow box half that size. You must remember when setting up a black light shadow box, that the paint only responds to the areas where the ultraviolet light can shine directly on it. The shadow box needs to be deep enough so the light will shine right on the models. The light may also be positioned outside the box. The color photograph of the two Star Trek ships was taken with only black light illumination and with the black light about two feet in front of the models.

If you must have an in flight shadow box diorama, I would suggest you use the smaller versions of the Star Trek vehicles that AMT offers in a three-in-one kit. Similar size metal models of the Star Wars vehicles are available from Kenner. The hobby shops that specialize in static aircraft and armor models will likely have a number of unusual spaceship designs in cast metal that are intended for fantasy wargames. Those models will also lend themselves nicely to a black light space diorama. The smaller vehicles can be used with a 24 x 30 inch box that is only 18 to 24 inches deep; a far more practical size for most homes!

Special Effects Illumination

There's no reason why you cannot illuminate the interior of your space ships, provide exterior navigational lights similar to those on conventional aircraft, or even go the whole route to duplicate the *Close Encounters of the Third Kind* type of light show space ship. The Star Trek Enterprise can be illuminated using small light bulbs available at most model railroad stores (figure 10-8). Stamped metal brackets were used to hold the batteries, but it really would have been easier to use one of the 9 volt transistor radio batteries with the dress snap type of connections or a penlight battery. Electronics hobby stores sell both the batteries and the snaps and

Fig. 10-8 Model railroad shops sell these "grain of wheat" bulbs in 12-volt ratings. Protect the plastic with a one inch piece of aluminum tubing wherever the bulbs must protrude through it.

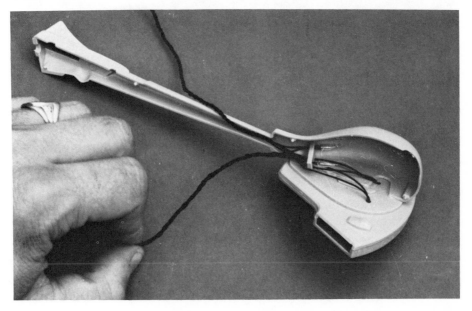

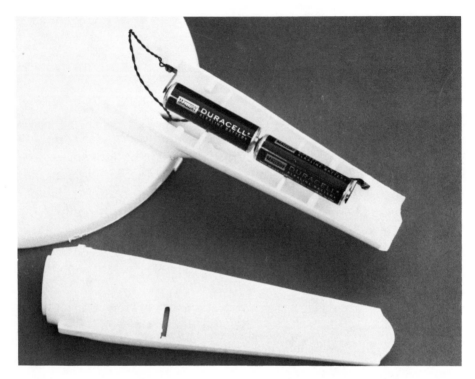

Fig. 10-9 A simple penlight would make a better battery holder than these metal clips. Just connect the wires to the empty bulb socket and to the case of the penlight and install the batteries.

they're perfect for the 12 volt model railroad bulbs because the lower voltage will double the life of the bulbs.

Be sure to line the interior of the model with aluminum foil if you're trying for interior lights. If you want exterior navigational or warning lights, I would suggest you cut off one inch bits of $\frac{1}{8}$-inch diameter aluminum tubing and insert them into hand-drilled holes in the model (use a pin vise and a $\frac{1}{8}$-inch drill bit). The aluminum should be reflective enough to keep the light bulbs' heat from melting the plastic. The excess length of tubing can be inside the model, of course, to provide the necessary mass for a heat sink.

Flashing lights are relatively simple with the Astable Multivibra-

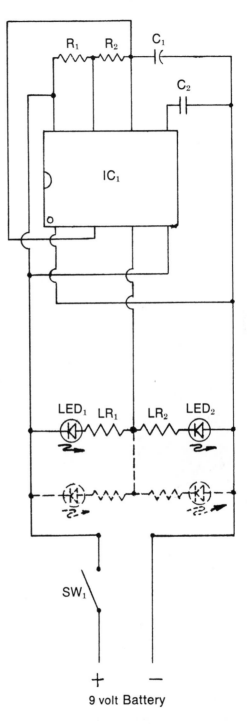

Key to Symbols

Symbol	Description of part
IC₁	555 timer
R₁	150K ohm ¼ watt resistor
R₂	100K ohm ¼ watt resistor
C₁	2.2mfd 10 volt electrolytic or tantalum capacitor
C₂	.01mfd 10 volt ceramic disc
LR₁	¼ watt resistor (see table for value)
LR₂	¼ watt resistor (see table for value)
LED₁	any green, red, or yellow light-emitting diode
LED₂	any green, red, or yellow light-emitting diode
SW₁	SPST (on-off) switch
Battery	9 volt transistor radio battery (Components will be damaged if battery voltage exceeds 12V or is less than 5V.)

LR₁ and LR₂ Values

LR Value	with X Number of LEDs
470 ohm	1
270 ohm	2
180 ohm	3
120 ohm	4
100 ohm	5
82 ohm	6

Fig. 10-10 The Astable Multivibrator circuit, from G R Signaling, will flash up to 12 red, green, or yellow light-emitting diodes (LEDs) for interior or navigational lights.

Fig. 10-11 Tape the LEDs and wires to a mock-up board to simulate their location inside your model so you are sure the lights flash where and when you want them. You can use either a transistor radio snap-top battery or a smoke alarm battery like this one.

tor circuit shown on these pages. The circuit will flash up to six light-emitting diodes (LEDs) on each side in a ping pong style cycle. A random number of the left set of six bulbs can be placed on the right surface of the vehicle to give a twinkle effect if you wish it.

The circuit is designed to utilize hobby components available at any electronics hobby store. Take the diagram with you and let the salesperson help you select the proper components. The light-emitting diodes appear to be tiny light bulbs (they're what make the numbers light up in most digital watches and clocks), and they're available in several sizes including a common .120-inch diameter and .200-inch diameter.

The LEDs are available in yellow, red, and green, but not white. If you use green LEDs, you'll have to put all of them on one side of the circuit because the green ones require more current than the yellow or red ones. If you try to mix red, green, and yellow on the same side of the circuit, only the red and yellow will flash on and off. In spite of this restriction, if the wires leading to each LED are long enough, you can still position any color in just about any place on the model.

So if the LED does not flash, check to see if there is a green LED in the circuit (or side) with red and yellow ones, or reverse the connections on the LED that does not flash.

Buy the plug-in style LED if you have a choice, because it takes an expert solderer to connect bare LED wires without burning up the LED itself. The wires can be soldered or wrapped to the LED sockets *before* the LEDs are plugged in to avoid that danger. A kit for this circuit is available for $10.95 (*less* any LEDs) from G R Signaling, 28504-93 Sand Canyon Road, Canyon Country, CA 91351.

You may also want to try using fiber optic lighting systems to place small pinpoints of light on the surface of the model as either portholes or navigational lights. The fiber optics use a length of flexible plastic rod .020 to .090 inch in diameter to transfer light to the blunt end of the rod. You can place hundreds of spots of light of just about any color on a model ETV by using a single penlight size light source and fiber optics. An experimental kit of fiber optics strands, a light source (less batteries), dyes to color the strands, and instructions are available by mail for $14.95 from Edmund Scientific, 1887 Edscorp Building, Barrington, NJ 18007. Some hobby or craft shops and some lighting supply stores may also have materials for fiber optics.

CHAPTER 11

Photography

The motion picture cameras used to film the special effects for the space fantasy films cost thousands of dollars each. Special dollies, lighting, film editing, and artwork on the film itself can cost millions. You can achieve almost the same results with still photographs and virtually no investment at all. You will need a camera that has an adjustable aperture and adjustable speeds to take photos like those you see on these pages, and you'll need a tripod to support the camera. If you're just experimenting, you can rent both the camera and the tripod.

Most of the pictures of vehicles in space you see in this book were taken of a model kit or a model made from kit and scratch components. Almost none of the photos are from any of the popular films or television shows. In other words, your photos can be just as interesting. The only secret can be revealed right now: it takes special lighting and longer than usual exposure times. The only photo taken with black light (ultraviolet) photography is figure 11-1. All the other photos were taken with conventional 3200K photoflood light bulbs or natural outdoor lighting.

Depth of Field

There are a few basic rules of model photography that you must learn and apply regardless of what you are shooting. The major difficulty presented by model photography is that you need to get fairly close to the subject, even the rather large ETV models. When you move in close with the camera, and focus on a particular subject, you lose what is called depth of field, and the background and

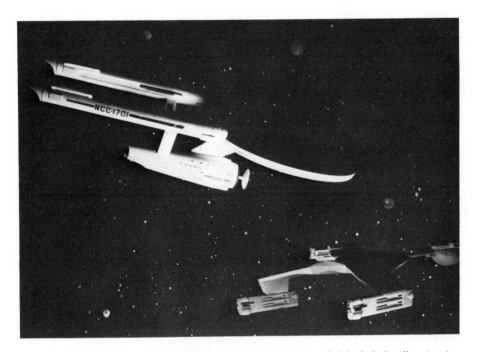

Fig. 11-1 This photo was taken with black light illumination in the shadow box/diorama in Chapter 10. The stars and the models were painted with ultraviolet-sensitive paints.

foreground become more and more blurred, the closer you get. There are a variety of close-up lenses, porta lenses, and lens extensions that will let you get as close as you want and still keep the subject in focus but there is no way to eliminate that depth of field problem. If you study some of the photos on these pages closely, you'll see that the farthest parts of some of the subjects (and also some of the nearest parts) are slightly blurred or out of focus. The trick is to almost fill the camera's lens with the subject even if you have to buy or rent one of the close-up lenses or adaptors to do it.

The best way to compensate for the limited depth of field in model photography is to use as small a lens opening (as *large* an aperture *number*) as you can. Most adjustable camera lenses will stop down to at least an f22. The close-up lenses like Nikon's Macro and

Fig. 11-2 Brighter illumination, from a photoflood bulb, reveals the corrugated carboard texture and even the nine-
pound test nylon fishing line used to suspend the models.

other brands of Micro lenses will stop down to f32. If you expect to
sell a photo of a model to a publication, you'll need a lens that will
stop down to at least f22. For your own hobby photography, an f16
is small enough.

The smaller apertures will restrict the amount of light getting to
the film. To get enough light for a picture, then, you'll either have
to use more light or keep the lens open longer than you normally
would and, more than likely, both. A tripod will be needed for any
exposure longer (slower) than about 1/60 of a second. To select your
exposure time, use the camera's built-in light meter, or buy or rent
a hand-held light meter, to determine how much exposure (light)
should be needed. I say "should" because the smaller apertures
usually need even longer exposure times than the meter will tell
you they do. Then "bracket" any photos you take by shooting one

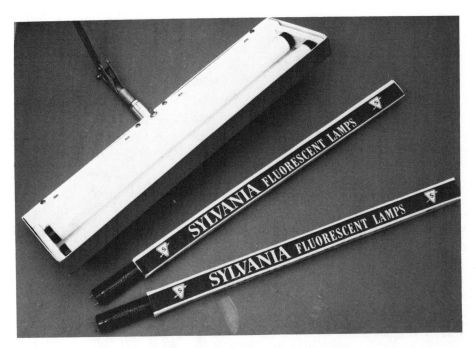

Fig. 11-3 Ultraviolet bulbs are available in tube styles to replace conventional fluorescent bulbs like these or in screw-in styles to replace incandescent bulbs.

at the smallest possible f stop with the exposure time suggested by the light meter, take a second shot with the next fastest exposure time, a third with the next slowest exposure time, and a fourth *three* exposure times greater than the meter would indicate.

For example, if the meter says that you should shoot an f22 stop with $1/25$ second exposure time, then shoot one shot at $1/25$, the second at $1/50$, the third $1/10$ (or $1/15$), and the fourth shot at a full half-second. Keep a record of all the readings so you can refer to it when the film is developed and prints or slides are made.

Black Light Photography

Black light photos take quite a bit of exposure time because they are indoor shots. There's no reason to shoot black and white with black light illumination, by the way, because the effect will be the same as with plain old fluorescent lighting. For color, I'd suggest using Kodak's ET Ektachrome for tungsten lighting. You'll have to go to a large camera store to find it, but there's no other Kodak film that will work as well for indoor model photography. The ET film has a relatively fast speed, but it will still require as much as 120 seconds of exposure time with an f22 stop!

You will have to experiment with filters to find the best one to suit your particular black lighting conditions; try a number 8 and a number 2B. The color photo of the Star Trek models was shot at f22 with 60 seconds of exposure time with a number 2B filter. Your cam-

Fig. 11-4 The Kenner Star Wars™ toy version of the T.I.E.-Fighter™ was taken with light bounced off a white ceiling and a blurred poster as a backdrop.

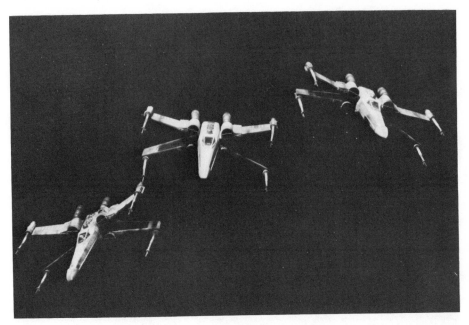

Fig. 11-5 These three MPC brand Star Wars™ X-Wing Fighters™ were photographed with light from two 3200K photoflood bulbs and reflectors held very close to the ground to help hide the shadows. The exposure time with Tri-X high-speed film was 120 seconds! (MPC kits are products of Fundimensions, CPG Products Corp.)

era and lens alone could change that to as little as 3 seconds or as much as 3 minutes. The only way to find out is to experiment with bracketed exposures.

In-Flight Photographs

The best way to make a model appear to be flying is to simply lay it on a piece of black velvet or a photo mural and light it so there are no visible shadows (to the camera!) That's how all the photos in this chapter (except the black light shot in figure 11-1) were taken. A single 3200K, 250-watt photoflood light bulb was placed in a conventional light socket with an aluminum reflector. The Kenner toy Star Wars T.I.E.- Fighter was then laid on a photomural of a racing scene

Fig. 11-6 Robert Angelo's MPC brand models of the Star Wars™ X-Wing Fighters™ were photographed on black velvet with rhinestones to simulate light reflecting planets. (MPC kits are products of Fundimensions, CPG Products Corp.)

with a blurred background. For the black and white photos (figure 11-4), the light was aimed at a white ceiling eight feet above the model to provide a "bounce" effect and all other lighting was turned off.

The Star Wars X-Wing Fighters, shown in figures 11-5 to 11-7, are MPC kits assembled and detailed by Robert Angelo. Those models were placed on a flat black background with about fifty rhinestones from dimestore jewelry to reflect the light like stars. To take photo 11-5, two of the 3200K photoflood bulbs and reflectors were placed on the ground and aimed at the models from about eight feet away. The low angle and the long distance combined with the black background to hide any shadows. Both color and black and

Fig. 11-7 A tripod, remote release cable, and a camera with adjustable f stops and exposure times are all that's needed for creating flying special effects stills of model ETVs.

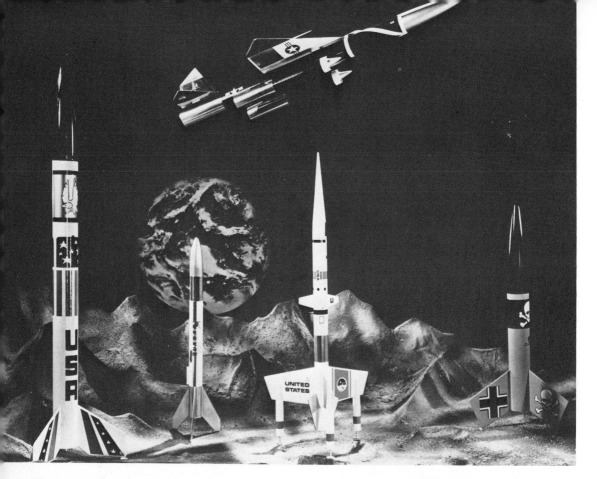

Fig. 11-8 A painted plaster lunar surface with a backdrop sprayed flat black and an earth cutout from a poster could be used to duplicate this photo. *Courtesy Estes Industries.*

white photos were taken with no other lighting. Kodak ET135 film was used for the color slides and Kodak Tri-X 135 for the black and white photos. The exposure times varied from about 30 seconds to almost 4 minutes using an f32 stop with a Nikon Macro lens.

It's much easier to photograph models outdoors with natural sunlight. The photo of the two X-Wing Fighters in figure 11-6 was taken on a piece of black imitation velvet with those rhinestones for stars. You should just be able to spot the shadows of the rocket models in the photo, but only because this shot was selected on the light side of a set of bracketed exposures to let you see the shadows. You won't find them in the other shots. You'll still need a tripod for the outdoor photos, but the exposure times should be more like fractions of a second ($1/100$ to $1/5$ or $1/2$, at most) than the one-minute or longer exposures needed for some indoor photos.

The only other trick I can suggest is a remote release cable for

174

Fig. 11-9 Two photos, one positioned on the backdrop to be slightly out of focus and the other supported just out of range of the camera lens, can produce this earth-to-moon effect. *Courtesy Estes Industries.*

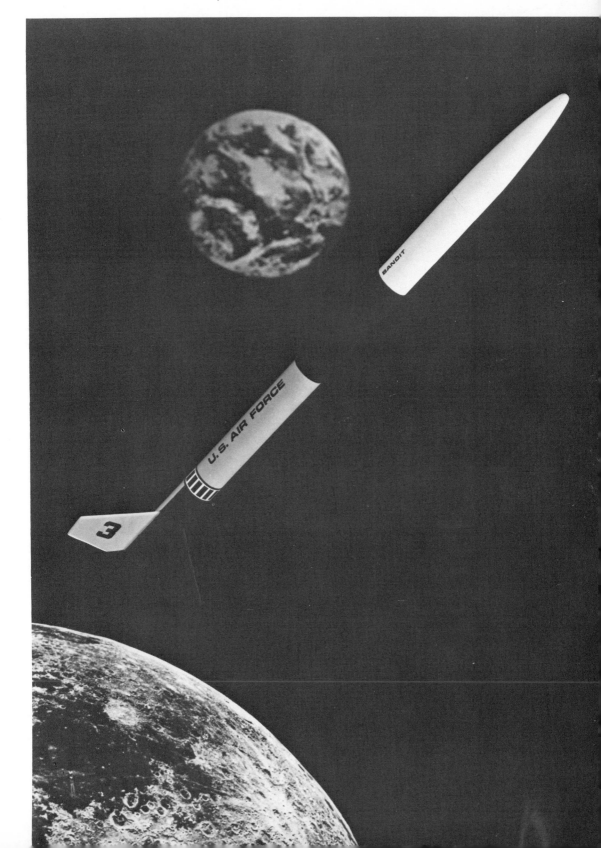

your lens shutter button. The cable will allow you to hold the lens button down for that minute or more without shaking the camera on its tripod. You will have to experiment with exposures by bracketing, even in black and white. I'd suggest you try the whole range from $1/100$ all the way to 5 *minutes,* skipping every other exposure time, for that first roll of black and white shots. Shoot at your camera's smallest opening, say f22, and at $1/100$, $1/25$, $1/10$, $1/2$, 2, 4, 10, 30, 60, 120 and 300 seconds. If you can, have your photo processor make a proof sheet or ask them *not* to compensate for exposure so you can see the results. When you get the system down pat with black and white film, you can try color.

Glossary

Air brush: An artist's tool that produces an atomized paint spray that can be adjusted for both pressure and paint volume. A miniature version of the spray guns used to paint automobiles.

Airfoil: A teardrop shaped cross section of a wing or other object that is designed to produce a lifting force perpendicular to its surface when there is motion between it and the surrounding air.

Apogee: The point in the flight of a rocket or in the orbit of a satellite when it is farthest from Earth.

Balsa wood: A light, porous wood grown in Ecuador that is used in the construction of flying model aircraft and rockets.

Booster: A portion of a rocket that is used to provide greater force or boost at lift off. The booster may or may not remain with the rocket for the rest of its flight.

Burnout: The point at which a rocket engine stops producing thrust.

Capillary action: The phenomenon of a fluid flowing on its own between two almost-touching surfaces or along a seam.

Card: Pressed paper stock that can vary from the .005-inch-thick single-ply to $1/16$-inch-thick cardboard. The term includes file cards, manila folders, and the varieties of cardboard mailing tubes used to construct the bodies of flying model rockets.

Center of gravity: The point at which a model rocket will balance if placed on a point. The point where the mass is distributed evenly in all directions.

Coast period: The portion of a model rocket's flight between burnout and the activation of the ejection charge. A delaying element in the rocket engine usually emits of trail of smoke to aid in tracking the rocket during this period.

Conversions: Utilizing parts from two or more model kits to produce a model that is different from any of the kits used for its construction. A

177

relatively easy method of duplicating a real rocket or creating a design not available as a kit.

Cyanoacrylate: A type of quick-setting cement that relies on the absence of air to cure.

Docking: The link-up of two or more space vehicles.

Ejection charge: A type of rocket energy built into most model rocket engines to eject a recovery parachute or to ignite the engine of an upper stage.

Epoxy: A two-part adhesive consisting of resin and catalyst that must be mixed. Epoxies "cure," rather than "dry."

Fin: The wing-like protuberances on a model rocket that serve to help guide its flight and provide directional stability.

Finishing: Usually used to describe the process of applying liquid grain-filler and paint to bare wood and smoothing each with sandpaper when dry, to produce a smooth exterior surface.

Igniter: The electrical device inserted into the exhaust port of a rocket engine. It is actuated by a remote source of current through a launch button with safety key.

Kit-bashing: See Conversions.

Kit-conversions: see Conversions.

Launch pad: The structure used to support a model rocket vertically just before and immediately after lift off or launch. The launch pad incorporates some type of rod or other device to guide the rocket for its first few feet of flight and usually serves as a place to house the alligator clips and wires that connect the rocket engine's igniter wires to the remote launch button.

Lift off: The first motion of a rocket upward after the engine ignites.

Missle: A propelled, unmanned vehicle.

Nose cone: The forward end of a rocket, usually made of balsa or lightweight plastic on a flying model rocket.

Nozzle: The exhaust duct in a model rocket engine that helps guide the exhaust from the rocket engine and to accelerate the exhaust gases.

Payload: The portion of a model rocket that can be removed to allow the insertion of lightweight instruments, biological specimens, an egg, etc. for research flights.

Recovery system: A means of allowing a rocket to return to earth undamaged for later flights.

Rocket: A thrust-producing system or a complete vehicle which derives its thrust from the ejection of hot gases generated from material carried in the system or vehicle.

Safety key: A special key needed to activate a model rocket launch so it will cause the rocket's engine to ignite.

Thrust: The force that propels a rocket, caused by the rearward ejection of the rocket engine's gases during the combustion process.

Zero gravity: The point at which weightlessness occurs.

Sources of Supply

AMT Corporation. See Lesney.

Aurora Products Corp.
633 Hope Street
Stamford, CT 06904

Badger Air Brush Co.
9201 Gage Avenue
Franklin Park, IL 60131

Ballantine Books
201 E. 50th St.
New York, NY 10022

Centuri Engineering Co., Inc.
1905 E. Indian School Rd.
Phoenix, AZ 85014

Champ (Champion) Decal Co.
P.O. Box 1178
Minot, ND 58701

L. M. Cox Manufacturing Co., Inc.
P.O. Box 476
Santa Ana, CA 92705

Estes Industries, Inc.
Penrose, CO 81240

Evergreen Scale Models
1717 N.E. 92nd Street
Seattle, WA 98115

Faller (Charles C. Merzbach Co.)
1107 Broadway
New York, NY 10010

Flight Systems, Inc.
9300 E. 68th St.
Raytown, MO 64133

Floquil-Polly S Corp
Route 30
North Amsterdam, NY 12010

Jerobee R/C Cars
12702 N.E. 124th St.
Kirkland, WA 98033

Kenner Products Division, CPG Products Corp.
1014 Vine St.
Cincinnati, OH 45202

Lesney Products Corp.
141 W. Commercial Ave.
Moonachie, NJ 07074

Lindberg Products, Inc.
8050 N. Monticello Ave.
Skokie, IL 60076

Liqu-A-Plate
Realistic Replicas
1097 No. State College
Anaheim, CA 92805

MPC Division of Fundimensions
CPG Products Corp.
26750 23 Mile Rd.
Mount Clemens, MI 48045

Micro-Scale
1821 E. Newport Circle
Santa Ana, CA 92705

Midwest Products Co.
400 S. Indiana St.
Hobart, IL 46342

Minicraft Models, Inc.
    1510 W. 228th St.
    Torrance, CA 90501
Monogram Models, Inc.
    8601 Waukegan Rd.
    Morton Grove, IL 60053
Mountains-In-Minutes
    (ISLE Laboratories)
    Box 173
    Sylvania, OH 43560
Pactra Industries, Inc.
    7060 Hollywood Blvd., Suite 101
    Los Angeles, CA 90028
PanaVise
    5224 Chakemco St.
    South Gate, CA 90280
Plastruct, Inc.
    1621 N. Indiana St.
    Los Angeles, CA 90063

Revell, Inc.
    4223 Glencoe Ave.
    Venice, CA 90291
Strombecker Corp.
    600 N. Pulaski Rd.
    Chicago, IL 60624
The Testor Corp.
    11500 Tennessee Ave.
    Los Angeles, CA 90064
Thayer & Chandler, Inc.
    442 No. Wells
    Chicago, IL 60610
Wm. K. Walthers, Inc.
    5601 W. Florist Ave.
    Milwaukee, WI 53218
X-Acto/CBS
    45–35 Van Dam St.
    Long Island City, NY 11101